About Face

An Artist's Guide to
Forensic Facial Approximation

by artist and writer
Michele Surcouf

About Face

An Artist's Guide to
Forensic Facial Approximation

ISBN 978-1-7385801-2-5 (paperback)
ISBN 978-1-7385801-8-7 (ebook)

International Edition

www.msurcouf.com

The images contained in this book have been exclusively created,
crafted, or photographed by the author, utilising drawing, painting,
sculpting, camera and computer image software to produce
illustrative content. Notably, any photographs of skulls featured
within this book depict replicas purchased from ©Bone Clones,
with express written permission obtained from ©Bone Clones for
the author's use of these images. It should be emphasised that the
purpose of this book is to supplement readers' knowledge and
skills and is not intended to serve as a substitute for scientific
research or study.

Acknowledgement

This book would not have been possible without the outstanding skulls from ©Bone Clones. Their expertise in creating accurate and detailed replicas of skeletal remains has been instrumental in bringing this project to life.

©Bone Clones' commitment to accuracy and attention to detail is evident in every replica skull they produce. Their replicas are meticulously crafted and ensure they are as faithful to the original specimen as possible. The quality of their work is a testament to their dedication to scientific accuracy and their commitment to helping scientists, educators, artists, and enthusiasts alike to understand human anatomy better.

As an artist, I have always been fascinated by the study of human anatomy and how the artist can use it to shed light on how we see the human body. But it is one thing to read about the skeletal remains and quite another to be able to examine them up close and personal.

The detailed replicas depicted in these images allow readers to explore the intricacies of each specimen and gain a deeper understanding of the subject matter. These replicas serve not only as a visual aid but also as an inspiration to those who seek to learn more about human anatomy.

I am grateful for the generosity of ©Bone Clones in allowing me to use the images of their replicas in this book. Their contribution has made it possible to provide readers with a unique and compelling visual representation of the fascinating subject matter contained within these pages.

In closing, I would like to commend ©Bone Clones for their unwavering commitment to accuracy and quality and their willingness to share their exceptional work with the world. I hope this book will serve as a fitting tribute to their fine craftsmanship and inspire others to delve deeper into the art and science of forensic facial approximation.

Visit ©Bone Clones at: https://boneclones.com

Forward by Javier Murcia

"I grew up learning from books like the one Michele has completed. In a fast world full of youtube videos and 3D models, this book has an old-fashioned taste that brings me back to my room when I was a teen looking to learn more about this sort of thing.

Michele balances the art and the science, putting into practice, step by step, all the information she has gathered in the field over the years. As fascinating as it is for those who like anatomical reconstruction, it is quite a challenge to get natural results; this is why I like Michele's work. The challenge fuels her to keep going and get results overcoming the obstacles she might encounter.

In my years as a professional sculptor, I have met very few people like her. She is a stream of positive energy looking to achieve that thing you are passionate about with endless devotion and making things happen; nothing stops her, which is why I admire her."

Javier Murcia, Sculptor

Javier Murcia, a Spanish sculptor born in 1981, studied fine arts at the University of La Laguna in the Canary Islands. Javier combined these studies with the study of anatomy at the Faculty of Medicine of the same university to develop a full command of this beautiful subject he is so passionate about. Visit his work at https://javiermurcia.com/

Table of Contents

Acknowledgement.. page 3
Forward... page 4
Author's note... page 7
What is forensic facial approximation?... page 9
The experts.. page 11
The artist or hobbies... page 12
We are born to see faces.. page 13
The artist's hand vs computer generation.. page 15
The bumpy road to identification.. page 17
Forensic facial approximation limitations... page 18
The artistry... page 19
Recognisability.. page 21
Methods.. page 24
Amplitude - bringing life to form.. page 26
The skull... page 27
Ethics... page 29
The anatomy of the skull.. page 30
Ancestry... page 33
Sex comparisons.. page 35
Materials... page 37
Sculpting tools... page 38
The clay.. page 40
Setting up the skull... page 42
Observational notes.. page 44
Finding muscle attachments. Origins and insertions.. page 46
Let's get sculpting / the reconstruction process.. page 49
Eyeballs.. page 50
Making Eyeballs.. page 51
Eyeball placement... page 54
Ears, making ears, ear placement.. page 55
Tissue depth markers.. page 58
Locations by number... page 59
Tissue depth locations description.. page 60

Nose... page 61
Muscles of the face... page 65
Joining the dots/ applying the clay.. page 67
Mouth and lips... page 69
Choosing an expression... page 71
Teeth.. page 73
The eyes... page 74
The eyelids and eyebrows... page 76
The nose in detail.. page 79
The finishing touches... page 80
Neck... page 83
Shoulders .... page 86
Hair... page 87
Refining your work .... page 89
Forensic facial reconstruction projects .... page 91
Elongated skull.. page 92
African American Female... page 100
African Male... page 109
Asian Female... page 118
Asian Male... page 127
European Female... page 137
European Male... page 146
Bibliography... page 156
About the Author.. page 158

Author's Note

Forensic facial reconstruction is a fascinating field that combines science, art, and history to bring the past to life. As an artist myself, I have always been intrigued by the human face and the stories it can tell. So I am excited to introduce this book on forensic facial reconstruction, now more often referred to as forensic facial approximation, which I hope will inspire amateur and professional artists and improve their understanding of the human face.

In this book, you will learn about the science behind facial reconstruction. You will also see how the science blends with the artistic skills required to create a facial reconstruction. Through step-by-step instructions and detailed illustrations, you will gain insight into the techniques used by forensic artists and how they bring a face to life from a skull.

This book has been written to inspire other artists to expand their skills and knowledge of the human face, whether they work in traditional media, digital art, or other forms of creative expression.

By learning more about the science of facial reconstruction, artists can better understand the anatomy and structure of the human face - from the inside out. This additional knowledge can be applied to any media, enhancing their ability to create realistic and emotive portraits.

I am honoured to share my passion for forensic facial reconstruction with you and hope this book will inspire you to explore this fascinating field and improve your understanding of the human face.

Michele Surcouf

What is forensic facial approximation?

"As poles to tents and walls to houses, so are bones to all living creatures, for other features naturally take their form from them and change with them."
Quote: Galen, (c. 129–199), Roman physician

Forensic facial reconstruction, or facial approximation, is the process of recreating an individual's face using their skull as a foundation. The technique is often used to identify unknown deceased individuals when other means of identification are unavailable or for historical and archaeological purposes. It combines scientific observation and artistry with sculpting, drawing, or computer graphics techniques to create the final result. Facial reconstruction is not an exact science but a marriage of scientific observation and artistry.

Facial approximation begins by placing tissue depth markers on the surface of the skull in precise locations. Clay is then placed over the skull, around and up to the top of the tissue depth markers, which determine the external boundary of the face. The "joining of the dots" from one marker to the next creates an approximate likeness of the individual and their features. Knowing how to sculpt facial features, with all their particular characteristics, is still essential for convincing results.

The primary goal of facial reconstruction is to identify a deceased individual and provide clues to investigators about their identity. The practice dates back to ancient Egypt, but modern forensic facial reconstruction began in the early 20th century when anthropologists and anatomists began using skull measurements to determine the characteristics of different populations. In 1935, Wilton M. Krogman, an American anthropologist, used a plaster cast of a skull to create a three-dimensional model of a person's face. Krogman's work became the foundation for modern forensic facial reconstruction.

Technology advancements have allowed for greater accuracy and consistency in the reconstruction process. In the 1970s, computer-based facial reconstruction techniques were developed, which involved creating a three-dimensional model of a skull using a computer. This model was then used to create a physical model of the person's face. Computer-based methods have since become more accessible and cost-effective.

Forensic facial reconstruction gained widespread attention in the 1990s when the practice was used to identify the remains of King Tutankhamen, the ancient Egyptian pharaoh. Since then, it has become a valuable tool in law enforcement and forensic investigations. It has been used to identify victims of crimes, including homicide and mass disasters, such as the September 11 terrorist attacks.

In recent years, advances in 3D printing technology have made facial reconstructions more accessible and cost-effective. Forensic facial reconstruction remains a crucial tool in forensic investigations, allowing investigators to identify unknown deceased individuals and solve crimes. While it is not an exact science, the combination of scientific observation and artistry has led to remarkable results, providing valuable clues to investigators and insight into the appearance of people from the past.

The experts

Forensic facial reconstruction is a specialised area of forensic science that involves using anatomical and artistic techniques to recreate the appearance of an individual's face based on skeletal remains or other evidence. If a forensic facial reconstruction expert is called to testify in court, they will typically need to demonstrate their qualifications and expertise.

The credentials required to testify as a professional forensic facial reconstruction witness may vary depending on the jurisdiction and the specific case. However, a forensic facial reconstruction expert should have extensive knowledge of human anatomy and facial structure.

In addition to their technical skills, a forensic facial reconstruction expert may also need a strong background in anthropology, archaeology, or other related fields. They may require advanced degrees in these areas and certification from professional organisations such as an Academy of Forensic Sciences.

Ultimately, whether an individual is qualified to testify as a professional forensic facial reconstruction witness will be up to the judge in each case. The judge will consider the expert's education, training, experience, and ability to explain complex scientific concepts to a lay audience.

The artist or hobbies

Forensic facial reconstruction requires a deep understanding of anatomy, facial features, and the ability to interpret and analyse skeletal remains. As a hobby, forensic facial reconstruction is not only fascinating but also extremely rewarding as it helps to bring back to life individuals who have been long gone.

One of the most exciting aspects of facial reconstruction is its challenge. Each new skull is a puzzle waiting to be solved, and the reconstruction process requires attention to detail and careful consideration of all the available evidence. It is a rewarding feeling to finally see the face of someone emerge.

Another aspect of this hobby is the opportunity to learn about history and culture. The process of facial reconstruction may involve research into the individual's life, the period in which they lived, and the social and cultural norms of their society. This hobby provides a unique window into the past and allows hobbyists to learn about different cultures and traditions.

Lastly, facial reconstruction offers a chance to use creativity and improve artistic skills. While the reconstruction process is based on science, an element of artistry is involved in creating a lifelike facial reconstruction. It requires a keen eye for detail and translating scientific data into a recognisable face.

Forensic facial reconstruction is an exciting and fascinating hobby combining science, art, and history. It offers a chance to solve puzzles, learn about different cultures and societies, and use creativity and improve artistic skills to bring the dead back to life.

We are born to see faces

As a 'face-focused' species, we tend to look for faces everywhere; in clouds, patterns, tree stumps, or even a piece of burnt toast. We even see a 'face' in something as simple as two dots and an arc within a circle.

Not only do we see a face in this symbol, but we also ascribe emotion to it. The phenomenon is called 'pareidolia' - the inclination to see a face where none exists.

While each individual's face is unique and highly complex, all faces require a fundamental similarity in their design. We are born to be sensitive to those unique and complex characteristics. Even the most identical of twins will have slight but discernible differences. It is these visible configurations of our facial features that allow others to identify us in an instant.

The face of an individual is what we tend to cherish most about those we have loved and lost. For thousands of years, we have venerated the memories of those deceased by painting, photographing, drawing, and sculpting their images. Through these methods, we 'reconstruct' a likeness that retains recognition of their face.

The desire to preserve the image (or, indeed, the actual head of the deceased) by reconstructing the face is an ancient practice. Evidence of this practice goes back to the Neolithic era.

During excavations conducted in 1953 in Jericho, two deposits were discovered beneath the flooring of houses located in the Pre-Pottery Neolithic "B" levels. Nine skulls were uncovered within these deposits, which had plaster built over the bone to form facial features. Additionally, shells were placed into the eye sockets to serve as eyes.

These ancient rituals may have been more of an attempt to preserve the person's lifelikeness, being, or 'soul', as opposed to how facial reconstruction is used today for identification.

Scientific attempts at forensic facial reconstruction gained popularity just over 120 years ago. In 1895, the renowned German anatomist, His, undertook a pioneering scientific investigation into identifying human remains. His specific focus was on the purported remains of Johann Sebastian Bach (1685-1750). To accomplish this, His gathered facial tissue measurements from a limited number of cadavers. Utilising this data, he created a bust which he then moulded onto a plaster cast of Bach's skull. Upon completion, the reconstruction was favourably compared to contemporary portraits and busts of the famous composer.

The artist's hand vs computer generation

There are essentially three popular methods of producing a facial reconstruction. There is 3D method, sculpting with wax or clay, 2D renderings (sketches), or computer graphic imagery that produces 2D or 3D computerised images.

Without the constraints of the scientific research data, the artist could use the skull as mere armature to construct any fanciful depiction they choose. Lack of adhering to the data could result in having little to no resemblance to the deceased.

Without the artistry, the scientific approach leaves the non-artist practitioner bereft of detailed information. Tissue depth markers determine tissue depth, but because depth markers represent a mean tissue depth pin-pointed on specific areas of the face, the non-artist may lack understanding of how the human facial features naturally flow into one another. Hence, the resulting image/sculpture can look strange and or lifeless.

Each method of facial reconstruction has advantages and drawbacks. While the artist (the sculptor or illustrator) may be able to finesse a lifelike reconstruction, often there is the issue of potential 'artist bias'.

The term "artist bias" refers to the recognisable style an artist uses in their work, which can make their artwork easily identifiable. This bias can also lead an artist to make certain features look similar. In facial reconstruction, artist bias may be almost unavoidable due to an artist's natural tendencies, but it is important to attempt to set aside these biases and follow the lead of the skull to create a more accurate representation.

No one's facial features are perfectly symmetrical on each side of their face. One eye might be slightly larger, or the nose may be slightly crooked, one ear might be higher than the other ear. It is these asymmetries that give someone their unique appearance. And these asymmetries are evident in the skull itself. These subtle asymmetries need to be taken into account by both the hands-on artist and the computer scientist to produce the characteristics that make someone remarkable.

While the last four decades have seen many advances in computer generated reconstructions, particularly in speed and accuracy, there are still some programme limitations. One of the limitations of some computer graphic facial reconstruction is the use of generic data. Databases lack all possible combinations of feature shapes, textures, etc.

To summarise, computer-based and artist-driven forensic facial reconstruction techniques can each produce good outcomes, but each has advantages and limitations.

Computer-based methods use 3D imaging technology and software to create a facial reconstruction. They are often more objective and accurate than traditional methods, as they use scientific data and measurements to determine the shape and dimensions of the skull and facial features, and they don't suffer from 'artist bias'. They can also be faster and more efficient than manual methods.

On the other hand, artist-driven methods rely on the skills and experience of a forensic artist, who creates a reconstruction by hand using traditional methods such as modelling clay. These techniques can provide a more nuanced and detailed representation of an individual's appearance, particularly concerning finer details like skin texture and wrinkles.

Ultimately, the choice between computer-based and artist-driven forensic facial reconstruction will depend on the specific case and the available resources. Both methods can produce accurate and compelling results, but one may be better suited for different cases or situations.

The bumpy road to identification

Producing a potentially identifiable face from only the skull is multifaceted. The reconstructive artist can reproduce some of the facial features with reasonable accuracy, while other characteristics can not be known with certainty.

When an unidentified skull is acquired, gathering as much archeological, ancestral, forensic, or historical information as possible helps create a better result. This information comes from anthropology, dentition, and law enforcement/forensics experts, answering questions such as: Where was the skull found? Could this determine ancestry (potential hair and eye colouring)? What was the approximate age of the deceased at the time of death? What is the sex of the individual? Was there any clothing found with the deceased that might determine physical stature? Was hair or artefacts found with the deceased? Were there any outstanding features, such as evidence of old injuries or disease, congenital disabilities or lifestyle impacts (e.g. unusual tooth wear)

This preliminary information assists the forensic artist in determining what the deceased may have looked like, and then this data is applied to the reconstruction. Vital information, such as the approximate age at the time of death, can give the artist clues about how the face may have aged. The artist can then apply general aged features to the reconstruction, such as deepened labial folds, crows-feet around the eyes, thinner lips, marionette lines around the mouth, loss of buccal fat in the cheeks, and heavier eyelids.

If a positive identification is not made after the facial reconstruction process, at the very least, it may rule out other individuals.

DNA sequencing has revolutionised the identification of deceased individuals. However, family members must be available to provide DNA samples to confirm a match, meaning there is still a place for hands-on forensic facial reconstructions.

Forensic facial approximation limitations

The facial approximation process has limitations regarding expression lines or scars that make the reconstructed skull more identifiable. Without information about the deceased or a photographic history, the artist cannot anticipate these features when reconstructing the skull.

Unless forensic evidence indicates signs of injury, the artist can not determine if there were any visible scars.

It is said, "we wear the face we make". The skin and soft tissues of the face are malleable and, with advancing age, will reflect the facial expressions we make most often when expressing emotion. Laughing, smiling, frowns, prolonged grief, squinting, and lip pursing are just some of the facial expressions that, over time, will cause the skin to form permanent lines and wrinkles.

Again, the artist can only anticipate expression lines when reconstructing the skull and must also rely on general information and observation on how the human face ages.

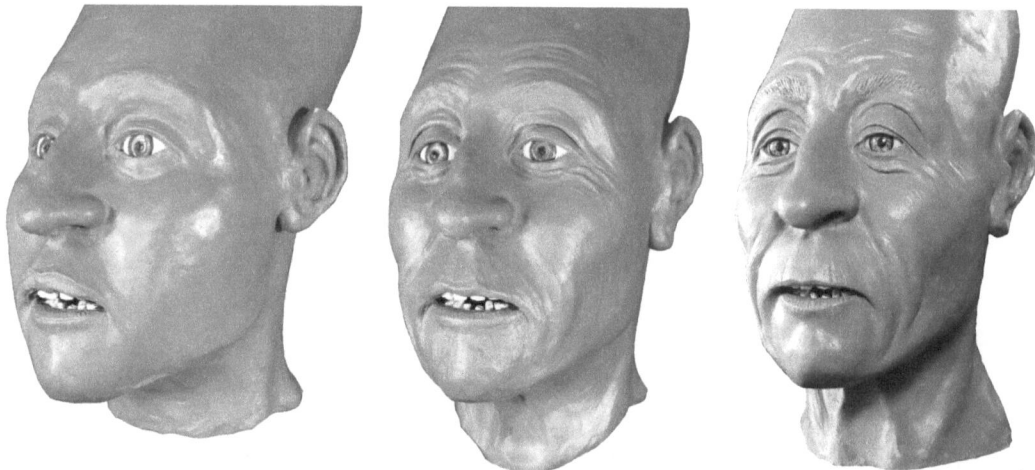

Deepened labial folds, crows-feet around the eyes, thinner lips, marionette lines around the mouth, loss of buccal fat in the cheeks, and heavier eyelids become more pronounced as we age.

The artistry

Can the forensic facial reconstruction artist execute a good likeness of a person by building the muscles and skin onto the skull - from the inside out?

Artists have studied the human face for centuries to depict it accurately in their artwork. A critical aspect of this study is understanding the ratios and placements of the various features of the face. These ratios and placements are essential for creating a realistic and anatomically correct portrayal of a human face.

One of the most well-known ratios artists use when drawing a human face is the Golden Ratio. This ratio is also known as the divine proportion or the golden mean, and it is a mathematical concept that appears in many aspects of art and nature. In the case of the human face, the Golden Ratio is used to determine the ideal proportions of the various features.

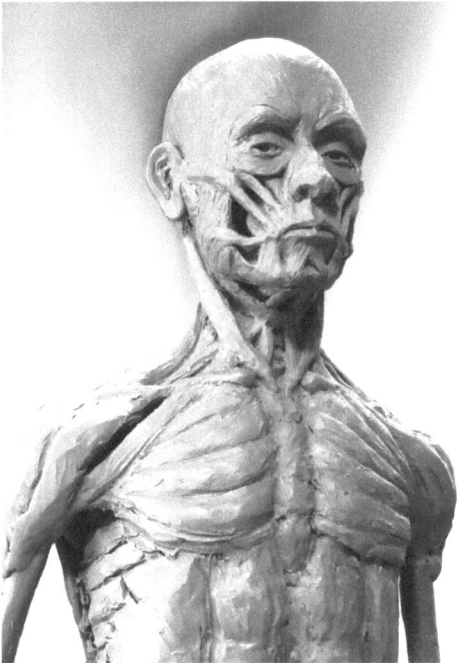

Is it possible to identify someone from muscle and skin alone? Écorché figure. Oil-base clay. 2021

The Golden Ratio dictates that the width of the face should be 1.6 times the height of the face. This means that the distance from the hairline to the eyebrows should be one-third of the height of the face, and the distance from the eyebrows to the tip of the nose should be one-third of the height of the face as well. The distance from the tip of the nose to the chin should be one-third of the face's height, resulting in the face being divided into thirds.

In addition to the Golden Ratio, artists use other ratios to create a human face. One such ratio is the nose's width compared to the face's width. Ideally, the width of the nose should be one-fifth of the width of the face.

19

Another important aspect of creating a realistic human face is understanding the placement of the various features. The eyes, for example, are typically located in the middle of the face, with the distance between them being roughly equal to the width of one eye. The eyebrows are typically located just above the eyes, with the highest point of the eyebrow being in line with the outer edge of the iris.

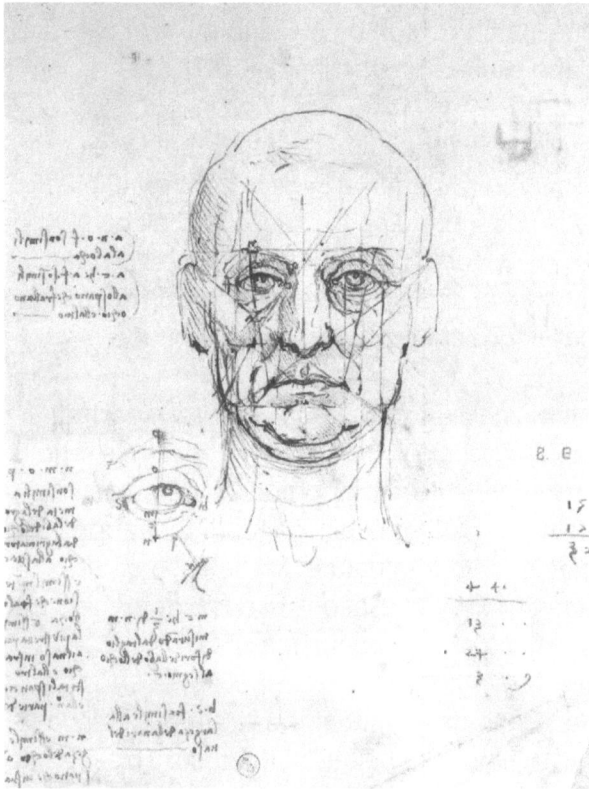

Leonardo da Vinci was a master of the arts and sciences, and his understanding of facial proportions was unparalleled. Leonardo's knowledge of facial proportions informed his artistic works and contributed to the field of anatomy and physiology.

The nose is typically located in the centre of the face, with the bottom of the nose lining up with the bottom of the ear lobes. The mouth is located slightly above the midpoint between the nose and the chin, with the corners of the mouth lining up with the pupils of the eyes.

Understanding these ratios and placements is helpful for creating a realistic and anatomically correct portrayal of a human face. Using these guidelines, artists can create portraits that accurately reflect the proportions and features of the human face.

Now that the ideal proportions for the human face used by artists has just been outlined, keep those rules in mind while reconstructing a skull.

However, everyone is slightly different. To create a face most faithfully, using forensic facial reconstruction, the artist must let the skull dictate where the features are located and let the skull give clues as to what those features look like.

It is the differences in characteristics that make people unique and recognisable.

Recognisability

The human eye is very discerning, sometimes noticing the most minute irregularities in shape, colouration, dimension, etc. And yet, we can view a very pixilated image and immediately know the individual's identity.

We can also look at a caricature of someone, where the features are grossly exaggerated, and again easily recognise the individual. We have also seen realistic representations (e.g. paintings) of someone that misses capturing the sitter's essence. So what is it that makes a face recognisable?

Image of Abraham Lincoln: pixilated and exaggerated features

For over a century, forensic reconstruction researchers have embarked on the quest for the perfect tissue depth average by using a variety of methods; taking measurements of recently deceased prior to embalming, after embalming, in a prone or sitting position, and even using CT scans of live subjects.

While having a good foundation of statistical data before embarking on a facial reconstruction is essential, is the exact tissue depth of the face relevant to recognisability? Has averaging tissue depth results hindered an identification in individual forensic reconstruction cases?

The lack of detail and a departure from accurate scale (as shown in the images of Abraham Lincoln) indicates that we have broad and multiple systems of what makes someone recognisable.

It could also be argued that we lose characteristics that make someone uniquely identifiable by averaging the data, and hence the features. Not only are the features such as overall shape, shading, interior features (eyes, nose, mouth), and colouration important in identification, but that caricature - the exaggeration of features that deviate from the norm - may be an essential part of how we recall and identify faces.

Instead of removing individualisation from reconstruction cases by using averages, could the application of caricature in forensic reconstruction improve recognisability - perhaps even more so than accuracy?

It is one thing to make a caricature of a face that is already known, so the overriding question for forensic reconstruction artists is; are the characteristics that make a face unique already implied in the skull? Can amplification of those characteristics, through degrees of characterisation, be used to improve identification? Unfortunately, these questions can not be answered without further study.

Nevertheless, completing a forensic facial reconstruction project, even just as practice, can benefit artists looking to improve their understanding of the human face and hone their skills.

Forensic facial reconstruction involves using scientific and artistic methods to recreate the appearance of an individual's face based on their skeletal remains. By participating in this process, artists can gain valuable insights into the structure and proportions of the human face and develop a deeper understanding of how facial features relate to one another.

Additionally, working on a forensic facial reconstruction project requires a high degree of attention to detail, which can help artists improve their accuracy and precision when drawing or painting portraits.

Overall, completing a forensic facial reconstruction can be an excellent way for artists to expand their knowledge and skills, regardless of whether or not they plan to pursue a career in forensic art.

Methods

Not having access to a learning institute where I could study hands-on forensic facial reconstruction, I decided to teach myself. I purchased reference books, watched YouTube videos, and downloaded various studies from the internet. The textbooks I most helpful and recommend for further information are listed in the back of the book.

Two facial reconstruction methods stood out. They are the Manchester Method and the American Method.

The Manchester method of facial reconstruction, also known as the Manchester approach, was developed in the early 1970s by Professor Richard Neave at the University of Manchester in England. Neave was a renowned forensic anthropologist, and his method aimed to improve the accuracy of facial reconstruction from skeletal remains, particularly in cases where only fragmentary or incomplete remains were available.

The Manchester approach employs a combination of anatomical knowledge, artistic skill, and scientific data to reconstruct the soft tissue of the face. Neave's pioneering work in this field has significantly impacted forensic science and has been used in numerous criminal investigations and historical reconstructions.

The American method of forensic facial reconstruction has its roots in the mid-twentieth century when researchers began exploring new techniques for identifying unknown individuals. One of the earliest pioneers of the method was Wilton M. Krogman, a physical anthropologist who developed a system of craniofacial superimposition in the 1940s. This technique overlapped photographs of a skull with images of a living person's face to create a possible likeness.

In the 1970s, Betty Pat Gatliff expanded on Krogman's work by developing a more detailed facial reconstruction process that relied on tissue depth markers and anatomical knowledge to create a more realistic likeness.

Which method to choose?

❋ The Manchester Method builds the muscles onto the skull before adding the fatty tissues and then the skin.

❋ In contrast, the American Method uses strips of clay directly on to the skull in a point-to-point reconstruction. The tissue depth markers are used as the guide.

With an interest in anatomy, I was initially drawn to the Manchester Method, building the face, starting with the underlying muscles, and then building the tissue up and out to the depth markers. This method ensures that the substrate of the face is in place and that the overall volume is correct, placing fat deposits and adding the skin, before finishing the face.

The American Method is faster and perfect for some areas with little to no underlying musculature. Whilst I have adopted this method in part, the concern is that without the underlying muscles, or a good understanding of facial anatomy and/or sculpting ability, the depth markers alone wouldn't be enough to ensure a good result.

Ultimately, I developed a facial reconstruction style that effectively blends the best of both methods while incorporating fundamental principles of figurative art.

Amplitude - bringing life to form

The human body, or any living body for that matter, has 'amplitude', meaning a body has fullness from within. This fullness is due to the presence of muscles and the inner pressure of bodily fluids within the body, particularly blood and the pressure it produces.

This internal pressure means there are virtually no true flat or concave areas on the surface of the human body. To depict a living person, it is essential to convey 'amplitude' in your facial reconstruction. By merging and blending one area of amplitude, or fullness, into the next area of fullness, the appearance of living flesh can achieved.

Each space between the tissue depth markers is depicted as a curved arc. However, each curve also possesses a unique characteristic - its amplitude or fullness, which is the extent to which the curve bulges out. Some curves are 'lean' with a flattened appearance, others are 'fat' with a full and generous shape, and many others lie in the middle.

Every shape extends and recedes. In artistic terms, each form emanates from a starting point, rises to its highest point, and then descends to another point of entry that becomes the starting point for the next curve that emerges.

The mouth area provides an excellent opportunity to witness the effects of amplitude in action. Various muscles converge here, producing numerous distinct curves. The lips exhibit fullness and roundness, with certain regions that appear more flattened.

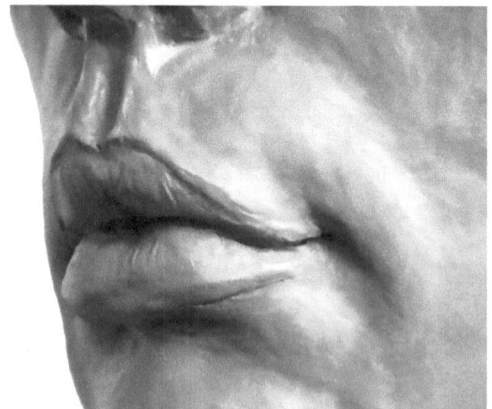

By meticulously observing and portraying amplitude, the lifelikeness of the reconstruction can be significantly improved.

The mouth displays 'amplitude'.

The skull

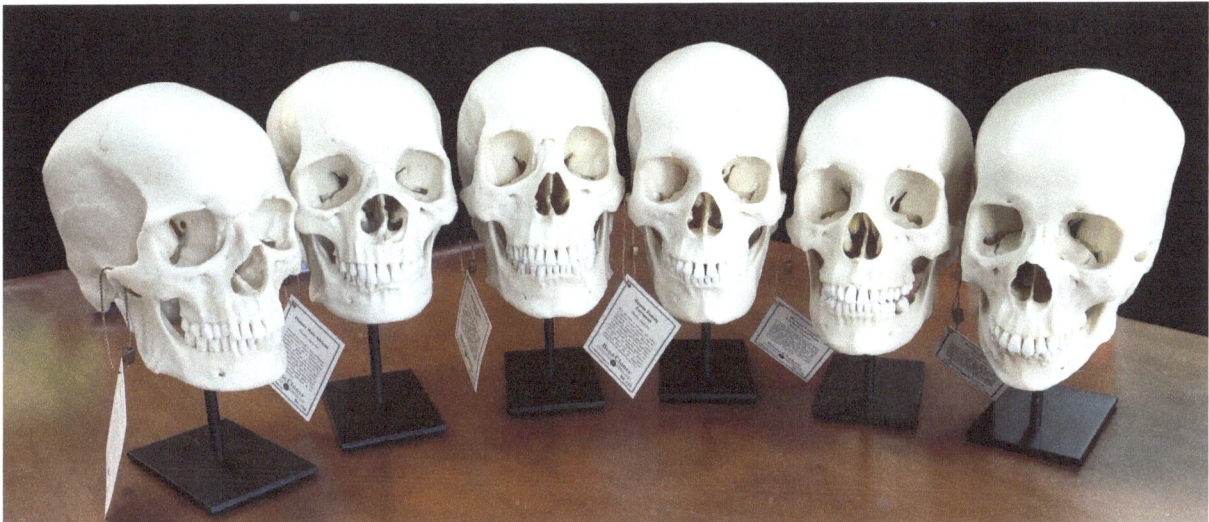

Replica human skulls can be purchased at ©Bone Clones, www.boneclones.com

To obtain a human skull, you need to work in a field such as forensics or archeology, where you can access human remains. Otherwise, finding a real human skull for practice can be difficult.

There are many issues with procuring a real human skull for training. Issues that must be addressed include the chain of custody and proper handling to avoid damage to the original skull. Still, the most critical issue is the sensitive issue of ethics.

To overcome the difficulties of obtaining a real human skull, I purchased replica skulls from ©Bone Clones in the USA. ©Bone Clones offers a wide range of high-quality and precisely detailed resin reproductions cast from real skulls. I highly recommend using these finely cast replicas.

The skulls used for this book have no photographic history, so there is no verification of this work being a faithful representation of the deceased. As in a real-life case, the artist must trust that the process will ultimately give the desired result.

The replica skulls used in this book are:

1. African American Female
2. African Male
3. Asian Female
4. Asian Male
5. European Female
6. European Male
7. 2000+yr old, elongated skull originating from Paracas, Peru

Replica human skulls can be purchased at ©Bone Clones, www.boneclones.com

Ethics

Human remains should always be treated with care and respect. Human bones of once-living people require ethical and legal consideration. The legal and ethical concerns regarding handling, storing, curating and analysing human skeletal remains vary from country to country. These considerations are grounded in specific historical and political contexts and are culturally meaningful to the descendants of the deceased.

Adhering to professional codes of ethics and standards is essential when using a real skull for forensic facial reconstruction. Please refer to your own country's legal standards for handling human remains.

The sale of human skeletal online is a lucrative black-market business. However, it is impossible to know if you are contributing to the black market practice of illegally acquiring human remains if you purchase skulls online from unknown sources.

Purchasing replica skulls through a reputable company like ©BoneClones, www.boneclones.com ensures compliance with legal and ethical concerns.

The anatomy of the skull

You don't need to know every bone in the skull to reconstruct it, but it is a good idea to be familiar with the names and locations. An explanation of the abbreviations below, and the exact points where the markers are placed, can be found in the chapter on tissue depths markers.

The head's complete bony structure is known as the skull. The lower jaw is referred to as the mandible. If the mandible is removed, the remaining part of the skull is called the cranium. The calvaria is the upper portion of the cranium, excluding the face. The whole facial skeleton is known as the splanchnocranium and includes the maxilla and mandible. Within this area, the most important facial features we associate with identity are located; the eyes, nose, and mouth.

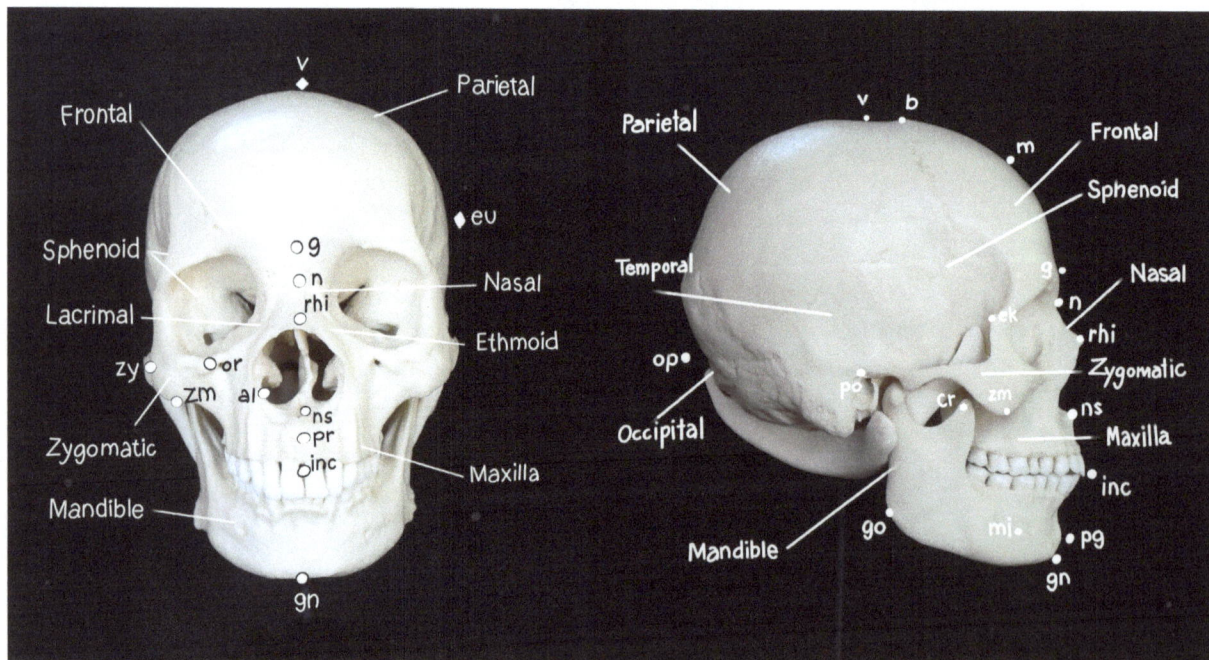

The bones of the skull are held together by sutures. Sutures are the fibrous joints between the bones of the skull that allow them to move and grow during development. In adults, these sutures become fused and immobile, creating a solid structure to protection the brain.

The appearance of the sutures varies depending on their location on the skull. Generally, sutures are irregular, interlocking lines of tissue where the edges of two bones meet. They are typically wavy and jagged in shape, with small, irregularly shaped bony protrusions called "sutural bones", like little islands, sometimes present within the suture.

Some of the significant sutures of the skull include the coronal suture, which runs from one side of the skull to the other and separates the frontal bone from the parietal bones, and the sagittal suture, which runs along the top of the skull and separates the two parietal bones. The lambdoid suture at the back of the skull separates the parietal bones from the occipital bone, while the squamous suture separates the temporal bone from the parietal bone.

Overall, sutures play an essential role in skull development and structure.

Skulls come in different shapes and sizes, categorised as dolichocephalic, mesaticephalic, or brachycephalic.

These classifications were first introduced in 1865.

In terms of facial characteristics, the 'dolichocephalic' face is characterised by eyes that are close together and set deep, a thinner and longer nose that protrudes more and is more likely to have a high root and aquiline profile, a forehead that

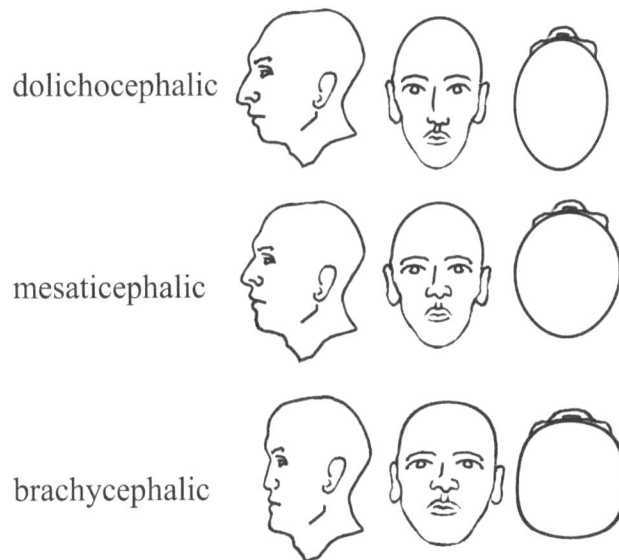

dolichocephalic

mesaticephalic

brachycephalic

slopes more, less prominent cheekbones, and a retrognathic facial profile, meaning the top teeth overlap the bottom teeth.

The term 'mesaticephalic' refers to a head that would be considered of average proportions, while the 'brachycephalic' face typically has eyes that are wide-set and exophthalmic, a broader and shorter nose with a more rounded nasal tip that is more likely to have a low root and concave profile, a more upright forehead, more prominent cheekbones, and a straighter or concave profile. This head type would be more prone to prognathism; and underbite.

Categorising the human head into one of three categories is important for several reasons. First and foremost, it helps identify and understand the differences in head shape and size among individuals from different ancestral groups. This knowledge can be useful in various fields, including forensic anthropology, medical research, and evolutionary biology.

Additionally, categorising the head shape can help identify certain genetic disorders or medical conditions. Overall, understanding and categorising head shape and size is a critical component of many disciplines and can have important implications for research and practical applications.

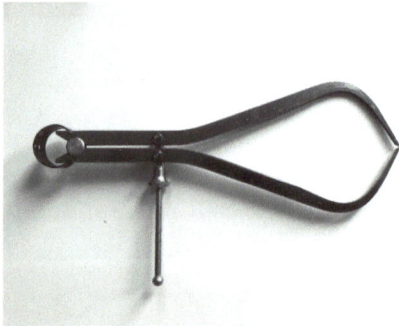

The calculation of the Cephalic Index (or cranial index) will determine which category the skull you are working on falls into. Use callipers (pictured) for these measurements.

Calculate the Cephalic Index by measuring the width of the skull (from side to side), dividing by the length (from front to back), and multiplying by 100. The result is a percentage that indicates if the skull is dolichocephalic, mesaticephalic, or brachycephalic.

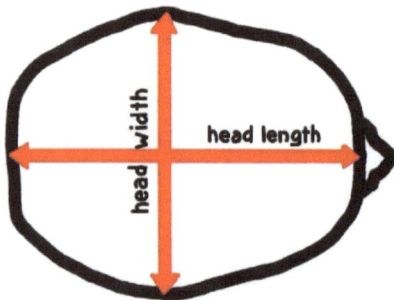

An index less than 75% is dolichocephalic.

An index of 75-80% is mesaticephalic.

An index more than 80% is brachycephalic.

$$\text{Cephalic index} = \frac{\text{head breadth}}{\text{head length}} \times 100$$

Ancestry

The human skull's shape, or morphology, is inherited from generation to generation. By analysing the skull in detail, certain features can indicate an individual's ancestry.

The following descriptions of the skull's morphology are generalisations, as there can be a lot of overlap in physical features. However, determining an individual's ancestry from identifying features can improve the chances of a more accurate reconstruction.

People of European ancestry typically exhibit facial features characterised by elongated, narrow faces with relatively flat profiles and sharply angled nasal bones. On the other hand, those with Asian ancestry tend to have broader and shorter faces with pronounced cheekbones. In contrast, individuals of African ancestry often have a wider nasal cavity and a prominent mouth area when viewed in profile.

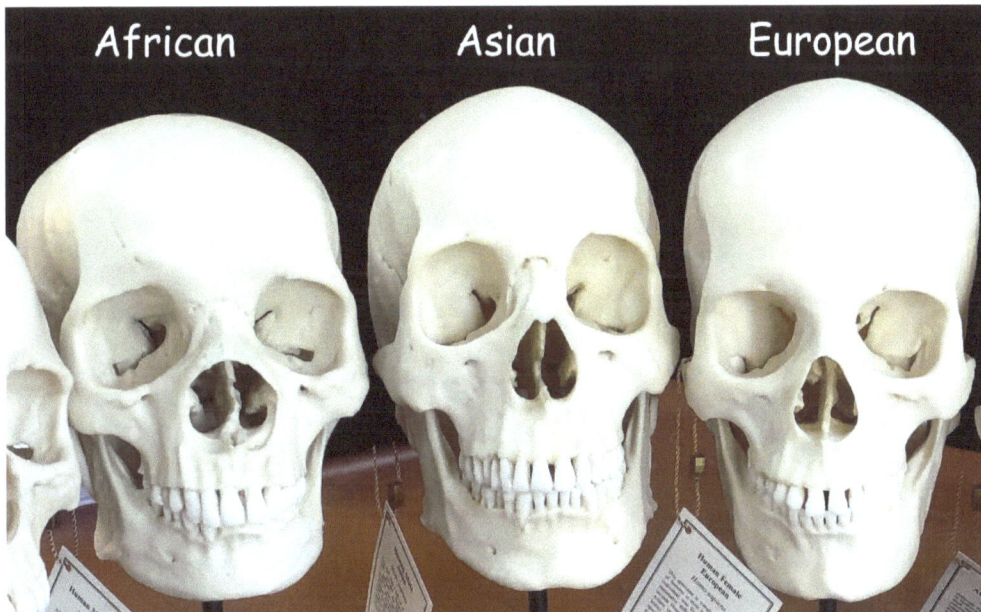

Comparing features of different ancestries.

Eye sockets:

* Those with European ancestry tend to be circular with squared margins, resembling 'aviator sunglasses.
* The eye sockets of those with Asian ancestry are circular.
* Those with African ancestry have rectangular eye sockets.

The nose:

* Those with African ancestry tend to have a broad nasal aperture and a flat nasal bridge.
* Those with European ancestry tend to have a more narrow, high nasal aperture and pronounced nasal bridge.
* Those with Asian ancestry tend to have a 'heart-shaped' nasal aperture and a less pronounced nasal bridge.

Jaw and teeth characteristics:

* Those with Asian ancestry tend to have 'shovel-shaped' upper incisors (teeth).
* Those with African ancestry tend to have larger teeth with wider spacing, and the jaw protrudes from the maxilla, called 'prognathism'.
* Those with European ancestry tend to have small teeth set closely together.

Sex comparisons

Distinct features of the skull can determine the biological sex of an individual. Below are comparisons of a biological male of African ancestry and a biological female of African/American ancestry. Note the differences between the two marked in red. In general, these differences will apply to any ancestry.

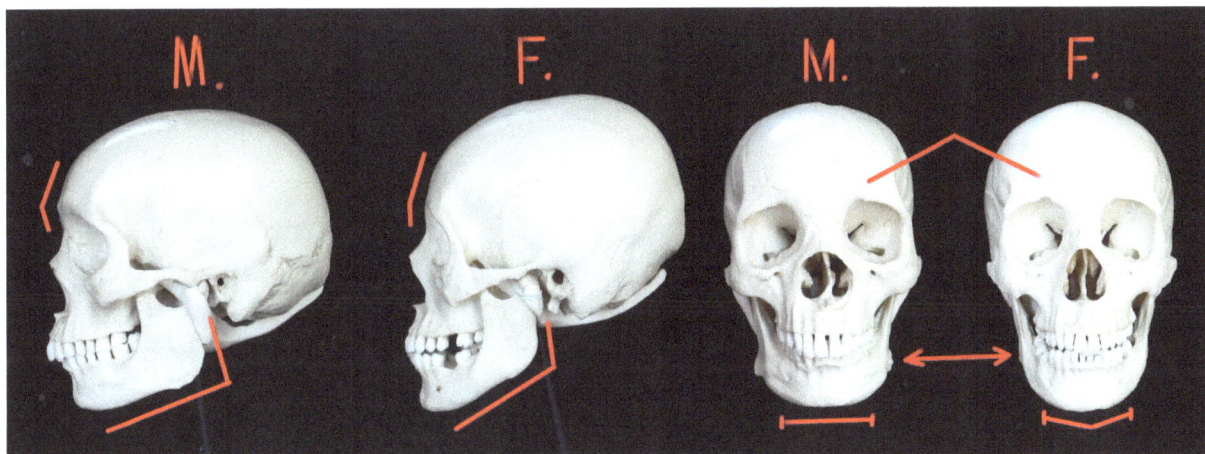

Comparing differences between a biologically male skull and a biologically female skull

Forehead and brow ridge:
Looking side on, the biological female skull has a more rounded forehead (frontal bone). The biological male's frontal bone is less rounded and tends to be more vertical. The bony ridge on the brow, above the supraorbital ridge (highest point above the eye socket), is more prominent in biological males and smoother in biological females.

Eye sockets:
Females tend to have a more rounded eye socket with sharper edges to the upper borders of the supraorbital. In contrast, male skulls have much squarer orbits with blunter upper eye margins.

Jaw:

Males then to have a square jawline at the gonion (the angle where the bottom of the jaw turns up toward the ear). This angle causes the line between the back edge of the jaw to sit more vertically in relation to the ear in biological males. In females, the jaw is more pointed at the front (the mental/menton), whereas the male jaw tends to be broader and square.

Typically, the skulls of males are characterised by greater weight, thicker bone density, and more distinct muscle attachment sites compared to those of females. Moreover, there are noticeable discrepancies in the shape of the forehead, eyes, and jawline that are utilised to establish the gender of a skull.

The female skull is generally:

* Smaller and lighter

* Rounded forehead

* Smooth supraorbital ridge

* Rounded eye sockets

* Sharp upper eye margins

* Pointed chin

* Sloping angle of the jaw

The male skull is generally:

* larger and heavier

* Sloping, less rounded forehead

* Prominent supraorbital ridge

* Squarer eye sockets

* Blunt upper eye margins

* Square chin

* Vertical angle of the jaw

The materials

Gathering the necessary tools and equipment to start your reconstruction is essential. You will need:

* ❄ The skull
* ❄ Clay
* ❄ Sculpting tools
* ❄ Measuring tools - callipers
* ❄ Some sort of tissue depth marker (short sticks, toothpicks or matches, glue gun sticks)
* ❄ Cutting tools - to cut the depth markers to length.
* ❄ Calculator

Callipers of some sort are an essential tool to make tissue depth markers correctly, take measurements for observational notes before beginning the sculpting process, and measure your work as you go. The yellow callipers (pictured) come with a probe that measures depths.

Measurement callipers, for tissue depths

Callipers can be purchased at hardware stores or homemade by simply cutting out two arches shaped like 'commas' (pictured below) in wood or heavy cardboard, drilling a hole at the top of each shape, and joining them together with a screw and wing nut.

Make the callipers so that they open and close. Ensure the callipers open wide enough to measure the length of a human head from front to back (approximately 30mm or 12 inches).

Homemade callipers

Sculpting tools

Sculpting tools are a vital component of creating lifelike reconstructions. However, it's about more than just using any tools but also knowing techniques to give your sculpture the desired texture and natural appearance.

The best tools for sculpting oil-based clay will depend on the specific area you are working on and the artist's preference. Ultimately, you will find the tools that work best with the clay you are using.

Tools that are commonly used for sculpting oil-based clay include:

* Metal sculpting tools: These come in various shapes and sizes and are perfect for creating larger shapes and smoothing out the surface of the clay.

* Wooden sculpting tools: These are great for creating organic shapes and adding texture to the clay.

* Modelling tools: These tools come in various shapes and sizes and are used for creating textures, patterns, and fine details on the surface of the clay.

* Dental tools: These are great for creating small, precise details and carving intricate shapes.

* Needle tools: These tools have a sharp needle point and are suitable for making precise cuts, incisions, and lines in the clay.

* Sculpting knives: These tools have a sharp, angled blade used for cutting and shaping larger areas of the clay.

* Scrapers: These tools have a flat or curved edge and are used for scraping and smoothing the surface of the clay.

* Wire loop tools: These are great for cutting and shaping clay and creating fine details and textures.

❋ Ribbon tools: These tools are similar to wire loop tools but have a flat ribbon shape. They are used for carving, smoothing rough areas, and shaping details on the surface of the clay.

❋ Clay shapers: These are made of flexible silicone and are great for creating smooth surfaces and blending one area of the clay into another.

❋ Wire brushes, toothbrushes, sandpaper, and pot scrubbers: These tools are great for adding texture and creating rough surfaces on the clay.

❋ Rolling pins: These can flatten the clay and create a consistent thickness throughout the sculpture.

❋ Hairdryers: These can help soften the clay, making it more pliable and easier to work with.

Most sculpting tools can be purchased at your local art supply store, hardware store, or online.

Below is a selection of sculpting tools, including metal tools, dental tools, wooden tools, and tools I have manufactured myself. More about tools and their uses are in the chapter on refining your work.

A variety of tools for sculpting

The Clay

Any professional-grade 'elastic' oil-based clay of medium hardness works well. It is suitable for heating and casting, good for manipulating when warm, and very hard when cold, so it holds carved detail well and can be handled easily. It also comes in several different colours.

A slow cooker or crockpot on a low heat setting is an excellent way to keep the clay at the right temperature. Speaking from experience, it is also a good idea to put the slow cooker on a timer so that you don't find a bubbling pot the next day should you forget to turn it off.

If you forget to turn off the slow cooker (and you find a bubbling pot of clay), it will have separated, the oil rising to the top and the clay sinking to the bottom. Just keep stirring it as it cools to recombine the clay and oil.

It may take some time to get used to this particular clay. Being hard when cold and runny when heated, you must find the right temperature for various applications.

Clay temperature tips:

✳ The clay must be hot enough to pour like a liquid when casting moulds.

Medium hardness oil-based heated in a kitchen crockpot - (© Monster Clay) https://www.monstermakers.com/monster-clay/

✳ The clay should be of medium heat to roll out with a rolling pin to create large pliable sheets of clay to cover large areas like the scalp.

✳ The clay should be only hand-warmed to apply to small amounts to areas of the face when filling imperfections.

✳ The clay should be cold to carve and hold finer detail.

✳ Clay can be warmed with a hairdryer for easier manipulation.

The clay will not harden over time, like water-based pottery clay, but it will remain firm as long as it is stored at room temperature and away from direct sunlight.

Setting up the skull

Before reconstruction, each skull must be set up correctly in the Frankfurt Plane. The Frankfurt Plane is the horizontal plane that intersects the inferior margin of the orbit (orbitale) and the external auditory meatus (porion) on both sides of the skull.

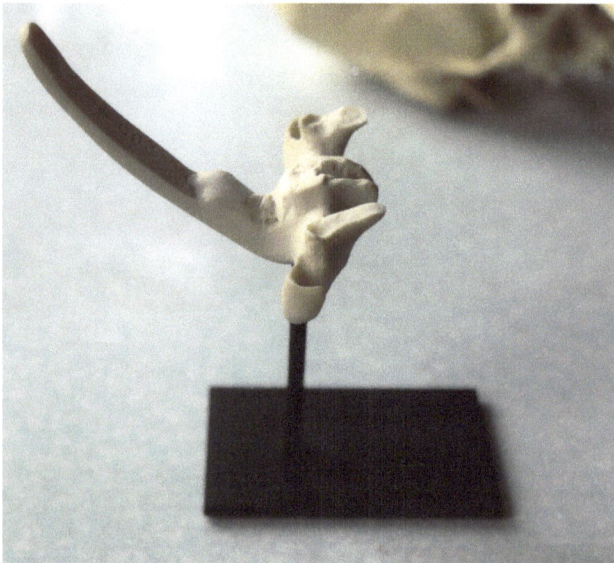

Stand for skull from ©Bone Clones

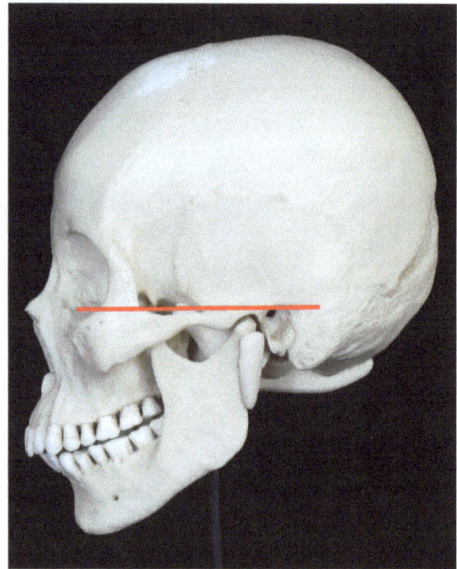

The Frankfurt Plane

In layperson's terms, the Frankfurt Plane is a horizontal line from the bottom of the eye socket and the ear opening.

Even though skulls from ©Bone Clones can be mounted on a stand (above right) and are an excellent foundation to build from, they only sometimes sit perfectly in the Frankfurt Plane.

In some instances, I manufactured a stand out of PVC piping. In other cases, I used a foam wedge under the accompanying stand to level the skulls into the Frankfurt Plane during reconstruction.

Manufacturing a separate stand with different materials is relatively simple. PVC piping is a cheap and easy way to create alternate stands. PVC can be cut to varying lengths with hacksaws, made malleable with a heat gun, drilled and securely screwed to a wooden base.

Different PVC pipe diameters and configurations (such as elbows and t-junctions) allow you to make various platforms to place the skull in its correct positioning and sleeves and inserts to remove your work easily. It is durable and easy to access at hardware and plumbing stores.

You can secure the skull to the PVC stand with warmed oil-base clay, which will become hardened and hold the skull in place as the clay cools.

You will be turning your work often to compare sides, so it is a good idea to set it on a turntable or place a cloth under your stand for easy turning, as small bits of clay can accumulate under the stand, making it sticky and difficult to turn. And also so you don't scratch your work surface.

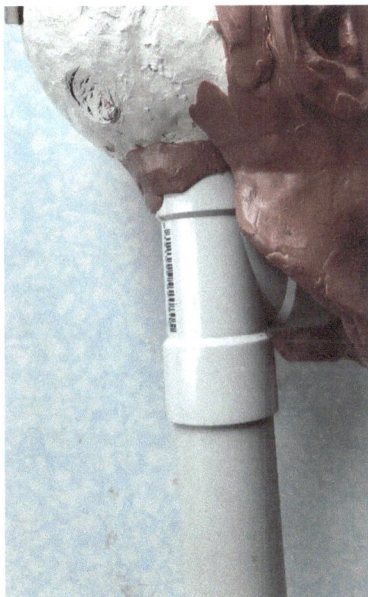

PVC piping for stand and elbow to rest the skull. Attached with clay.

PVC sleeve - cut with hacksaw, soften with heat gun, drill holes and anchor to wooden base with screws.

Observational notes

Once you have set up the skull in the Frankfurt Plane, making observational notes is the next and essential step before starting any reconstruction work.

* Observational notes allow you to see details that will give clues to the individual's physical appearance at the time of death, such as age, injuries, or birth abnormalities.

* Observational notes allow you to carefully study the skull and notice its particular peculiarities, asymmetries, and hints towards superficial features, which you will then go on to sculpt.

Skull purchased from ©Bone Clones come with an Osteological Evaluation prepared by Evan Matshes, BSc, MD Consultant Osteologist.

The Osteological Evaluation gives details of the deceased based on observations of the skull itself. However, I'd encourage you to do your own written observations before reading any reports. After completing your own written observations, read the official report and add any information from the Osteological Evaluation you've missed.

Careful observation of the skull will allow you to find where the muscles attach to the bone. Muscles attachments to the bone can be seen as roughness on the surface of the bone and can indicate how pronounced those muscles are. Close observation of the bone's texture can even allow you to see where the forehead ends and the hairline begins.

Without any further forensic or archeological information about the individual skulls, such as BMI, I treated them as if they were all of an average weight perimortem - the time of or just before death.

After completing the observational report for each skull, find as many tissue depth studies as possible (online studies, other resource books listed, etc. The studies will indicate ancestry, sex, body weight and age at death. Not all studies are the same, sometimes missing points on the skull and using different terminology. Where there are gaps in the information, you may have to use 'the next best' data and average it on a spreadsheet.

Observational notes should include:

1. General observation - ancestry, estimated age, sex
2. Overall skull shape
3. Forehead
4. Brow shape
5. Eye fissure
6. Eyeball
7. Eyelid pattern
8. Eyebrow pattern
9. Nasal Profile
10. Nasal width
11. Nasal spine
12. Alar shape and position
13. Nostril position
14. Teeth
15. Lip thickness
16. Philtrum width
17. Mouth width
18. Mouth corner inclinations
19. Lip shape
20. Nasolabial fold
21. Chin shape
22. Ear
23. Cheek shape
24. References

Finding muscle attachments: Origins and insertions

Muscles are complex structures responsible for generating force and movement in the body. They are attached to bones, which are connected through their origins and insertions.

The distinction between the origin and insertion of a muscle is important because it allows for precise and controlled movement and the ability to generate force in a specific direction. Understanding muscles' origins and insertions is essential for understanding human movement and anatomy.

✳ A muscle's 'origin' is the location where the muscle attaches to bone and remains immobile. It is 'anchored' to the bone when the muscle is in action.

✳ The muscle's 'insertion' is where its end, which attaches to another bone or tissue, moves during its action.

For example, the origin of the temporalis muscle along the side of the head at the temple feeds down behind the zygomatic arch (cheekbone), inserting itself into the top of the mandible (jaw. bone) at the coronoid process. The coronoid process is labelled 'CR' on the image of bones of the skull in the chapter entitled 'The anatomy of the skull'. The coronoid process is the forward-most prominent 'arch' of bone at the top of the jaw bone (mandible). This origin/insertion relationship is why you can feel the temporalis muscle flex at the side of your head (at the temple) when you clench your jaw closed or chew food.

We think of the skull as the scaffold of the face, but bone isn't the only determinator of how the muscles appear. Muscles can change the shape and density of bone. The discovery that ancient archers grew stronger bones in the arm that pulled back the bow is a fascinating look at this phenomena in action.

In the early 1900s, an archaeologist named Sir Leonard Woolley was excavating the ancient city of Ur in Mesopotamia (modern-day Iraq). Woolley discovered a tomb containing the remains of a man buried with his bow and arrows. It was thought that the man had lived around 2500 BCE during the Sumerian civilisation.

Woolley noticed something unusual about the bones of the man's left arm, which was the arm he would have used to pull back the bowstring. The bones in the arm were much denser and stronger than those in his right arm. Woolley hypothesised that this was due to the man's lifelong use of a bow and arrow.

Over the next several decades, other archaeologists and anthropologists began to study the bones of ancient archers worldwide. They found that the phenomenon of stronger bones in the arm that pulled back the bow was not unique to the man from Ur. It was a consistent pattern that could be seen in the remains of archers from many different cultures and periods.

Scientists then began to study the biomechanics of archery to try to understand why this was happening. They discovered that pulling back a bowstring requires a great deal of force, which puts stress on the bones in the arm. Over time, this stress stimulates the growth of new bone tissue, making the bones denser and stronger. Every time an archer drew the bowstring, the tension caused micro-fractures in the bones, which then healed and strengthened over time. This process is known as "Wolff's law," which states that bone will adapt to the loads under which it is placed.

As a result, archers who trained regularly and used their bow arm frequently developed denser and stronger bones in that arm. The strength differential between an archer's right and left arm was not limited to the bones alone but extended to the muscles and tendons, which also adapted to archery's demands.

This phenomenon is known as the "archer's paradox," and it is still being studied by scientists and historians alike. It is a fascinating example of how the human body can adapt to the demands of different activities.

In short, the stronger the muscle contractions, the more it pulls on the bone. This constant stimulation results in the bone producing more bone mass. This bone growth shows up on the skull as rougher surfaced patches, thicker, and even areas of flared bone. This increase in bone is particularly evident at the attachment points of the more powerful muscle, such as the masseter, with its origin at the bottom of the zygomatic arch (cheekbone) and its insertion at the base of the mandible (jaw).

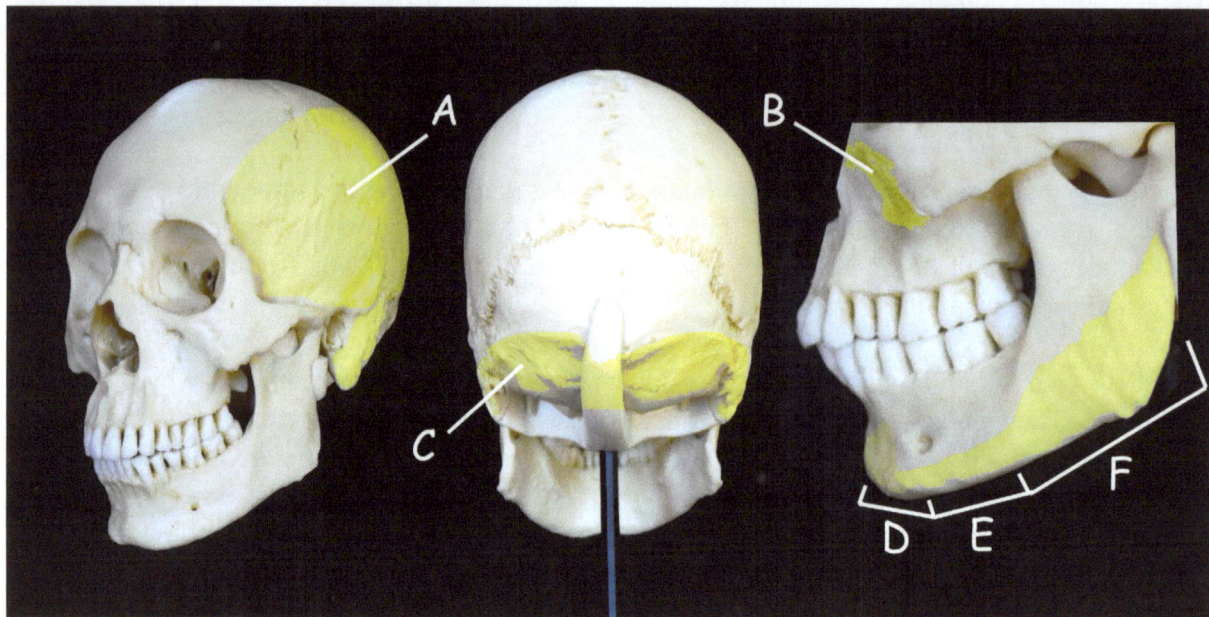

A rough surface indicates areas of muscle origins and insertions (highlighted yellow).

A. Temporalis (origin)

B. (origin) for Levator anguli oris, Levator labii superioris, Depressor anguli oris, Zygomatic minor, Zygomatic major

C. Trapezius (origin) and Sternocleidomastoideus (insertion)

D. Mentalis (origin)

E. Quadratus labii inferioris and Triangularis (origin)

F. Masseter (origin)

Let's get sculpting / reconstruction process

Initially I adhered to the step-by-step stages outlined in the reference books I consulted.

However, as I continued with these reconstructions, I evolved my own methods by integrating classical artistic observations and techniques, as well as the information I acquired from the research materials.

One such divergence was to put the scalp on first. Using the average tissue depth for each individual, I rolled out large sheets of the warmed and malleable clay and covered the scalp.

Covering such a large area gave an instant impression of bringing some life into the reconstruction project immediately.

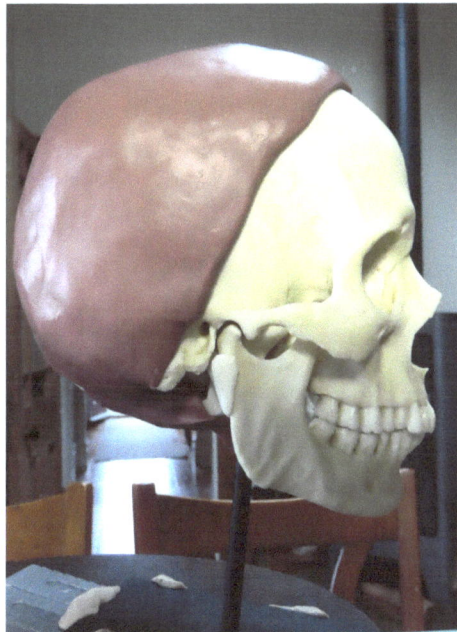

Adding the scalp

Eyeballs

Choosing the right eyeball colour and size will help make a face come alive and increase the potential of achieving an identification. Choose the size and colour based on ancestry, or the assumption of ancestry, if the eye colour is unknown.

There are several ways to acquire eyeballs.

1. Purchase glass, plastic or acrylic eyeballs of different sizes and colours from hobby stores or order online.
2. Sculpt the illusion of an eye by carving out the iris and pupil from the clay.
3. Craft your own with oven-bake polymer-type clay, inserting a colour for the iris and either use black clay for the pupil or a hole to capture the shadow, like the pupil of a human eye.

An adult human eyeball measures approximately 24 mm with no significant difference between sexes and age groups but can vary from 21 mm to 27 mm. So the exact size of the eyeball will be challenging to establish. The coloured part of an eyeball, or the iris, measures around 10 mm to 13 mm on average.

Counterintuitively, eyeball size doesn't correspond to eye socket size or depth. Asian ancestry averages a larger eyeball socket, but the eyeball has a smaller average than those with European or African ancestry. African ancestry has a smaller eyeball socket but the largest-sized eyeball on average.

The iris stands proud of the eyeball itself. The iris is still seen as slightly raised mounds under the eyelids when the eyes are closed.

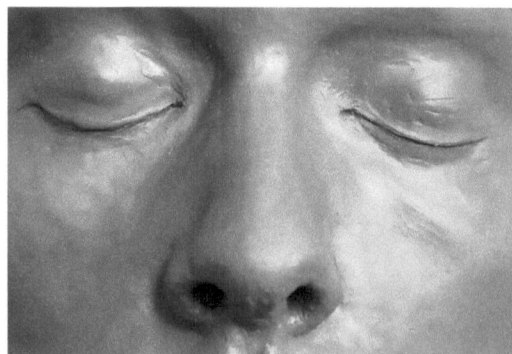

The iris under the eyelid

Making eyeballs

Make homemade eyeballs with a hobby polymer clay designed to be oven-baked. Whilst they may not be perfect, they do suffice for this project. The clay comes in many colours and can be mixed to create various colours.

Roll out the white clay with a kitchen pastry rolling pin to a thickness of around 3 or 4mm. Then press the clay into a form which is a half circle. A rounded measuring teaspoon works well. The white clay forms the sclera (the white part of the eyeball).

Then cut out a hole for the iris. I fashioned a tool from the mid-section of a fat ballpoint pen with a diameter of approximately 10 mm to cut out a hole, like a cookie cutter, in the white clay, where I placed the coloured iris. If the eyeball opening differs from the size of the iris, you can manipulate the soft clay to the extent you want.

Use the same tool to cut out the irises from the coloured clay. I used a small plastic straw to punch out the pupil in the centre of the iris and left it as a hole, much like an actual pupil's.

Eyes made from a bake-able coloured polymer clay

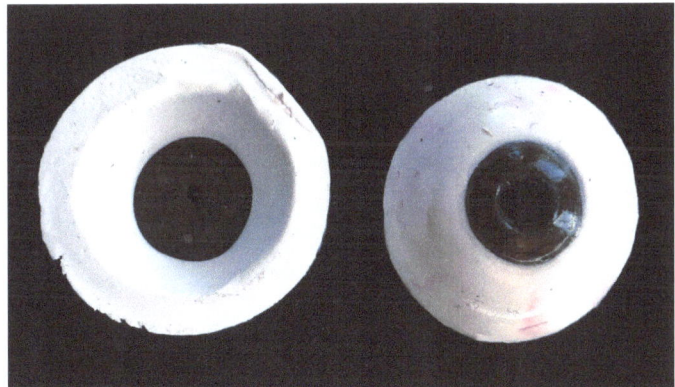

A measuring teaspoon-sized eyeball with coloured iris and pupil

51

* Bake the irises first and separately so that they are hardened when you place them in the hole of the eyeball itself. That way, you can manipulate the soft 'sclera' (white clay) around the iris for a good fit. Note that the iris sits proud of the sclera itself.

* Use different coloured clay, adding darker or lighter streaks through the iris to give it a more variegated look.

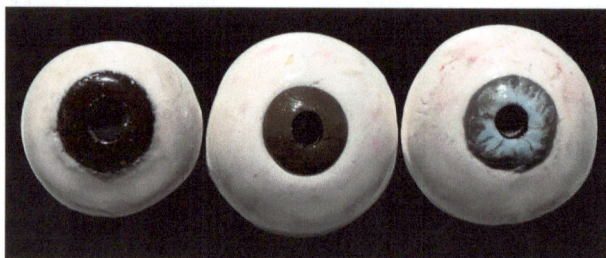

Various iris colours of and various sized eyeballs

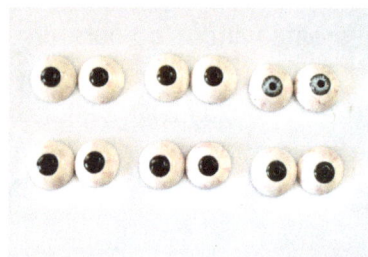

Selection of eye colours

* Bake the assembled clay 'eyeball' in the oven as directed by the manufacturer to hardness, then paint a glaze on the eyeballs to give the impression of being wet and shiny.

* Eyeballs are rarely completely white. Use a red pen and drew a few 'veins' on the sclera (white part of the eyeball) for a more realistic affect.

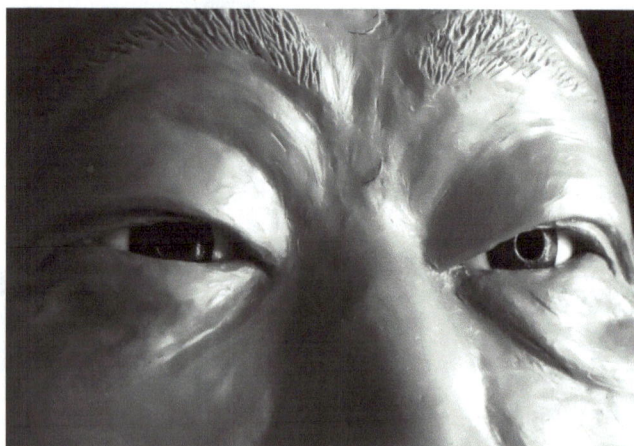

Polymer clay eyeballs - homemade

Plastic or acrylic eyes are a good way of depicting a realistic eyeball.

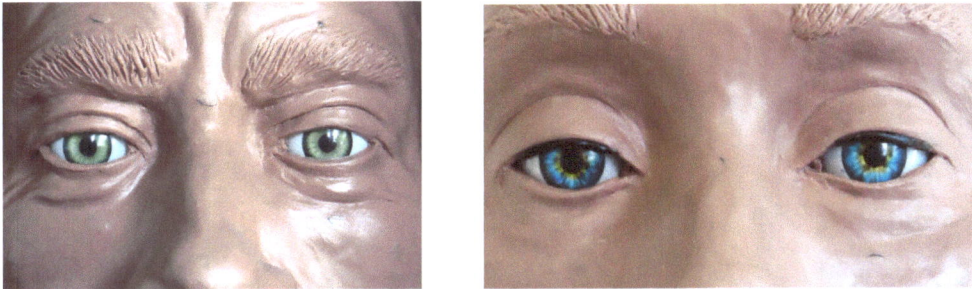

Acrylic eyeballs give a realistic look

✳ A small strip of dark wax placed just under the upper eyelids accentuates the eye, giving the impression of eyelashes (righthand image above).

Carving the iris and pupil from the clay itself can effectively create the illusion of an eyeball. The deeper the iris, the darker the eyes appear to be. So if you are trying to convey brown eyes, the iris would be carved deeper than it would for blue eyes.

Position the eyeballs so that the pupils face the same direction and appear to be focused on a point in space.

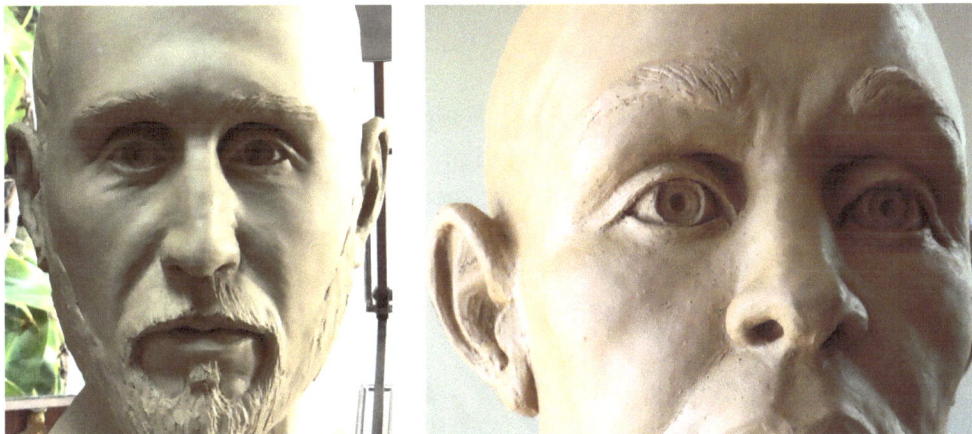

A sculpted eyeball, carved iris and pupil. Shadowing gives the impression of an eye.

53

Eyeball placement

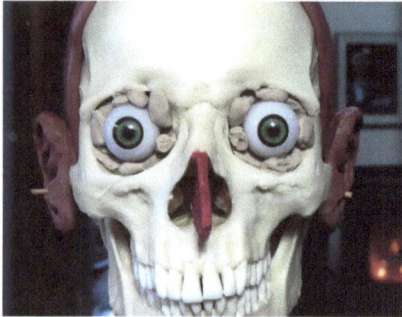

Eyeball placement - noting asymmetries.

Contrary to the sequence in other reference books, I preferred to place the eyeballs into the orbits at the beginning of the process.

This deviation allowed me to apply the surrounding tissue depth markers afterwards so that I did not knock the tissue depth markers off during the placement of the eyes. Asymmetries are often noticed at this point, as the orbit socket can vary in size and position from each other.

The process involves using soft and sticky oil-based clay. This clay is used to help position the eyeball in the correct location within the eye socket (also known as the orbit) and allows you to make minor adjustments to the eyeball.

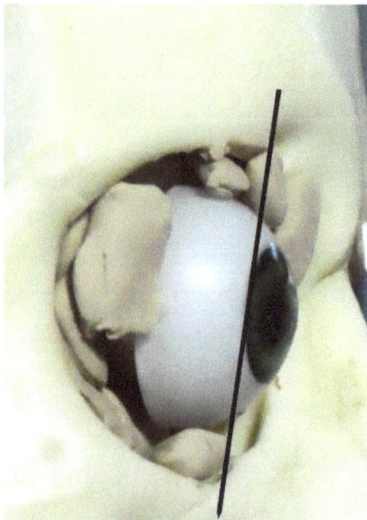

Eyeball depth

The clay is then moulded and shaped around the eyeball until it securely holds it in place. The goal is to centre the eyeball vertically and horizontally within the orbit.

When observed from a frontal view, the highest point of the cornea is located at the intersection of two lines. One line extends from the medial edge of the orbit (known as the maxillofrontale) to the outermost point of the nostril's edge (referred to as the ectoconchion). The other line bisects the orbit midway between the top and bottom edges.

The eye depth, or lateral (side) view is achieved by drawing an imaginary line from the superior and inferior margins of the orbit that runs through the outer point of the coloured iris (cornea).

Ears, making ears, ear placement

Human ears come in all shapes and sizes, and their positioning on the head can also vary. The shape of the outer ear can be round, oval, or even pointed, and its size and contour can affect how sound is captured and processed. The position of the ears on the head can also differ, with some people having ears that sit higher or lower than others.

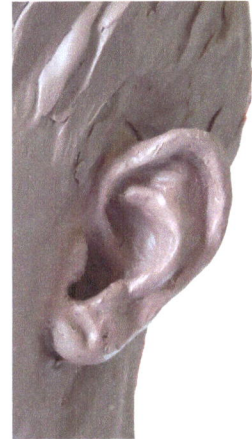

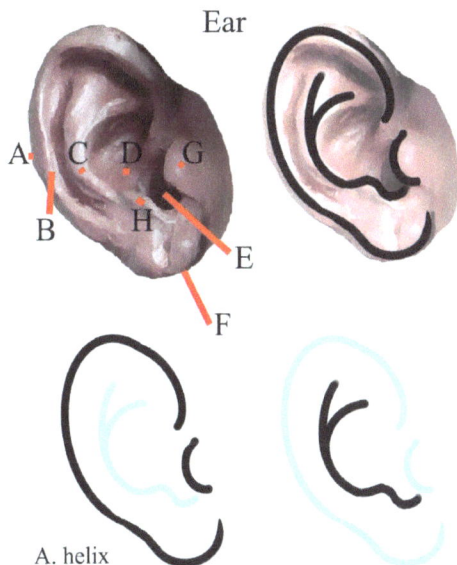

Ear

A. helix
B. helical margin
C. antihelix
D. conchal fossa
E. external meatus
F. lobe
G. tragus
H. antitragus

Ears can be daunting for the inexperienced artist. Initially, they look very complicated to draw or sculpt. There are a few tricks to understanding how to fashion the ear.

Consider the ear in two parts:

1.The bulk of the ear is somewhat like an upside-down pear in shape with a depression (conchal fossa) in the centre leading to the ear hole (external meatus). The edges on the top of the ears are called the 'helix' and fold toward the centre, creating a lip (helical margin). The bottom of the ear is rounded and fleshy, creating the 'lobe' of the ear.

2. A 'Y' shaped mound of cartilage (antihelix) sits on top of the 'conchal fossa' and within the depression of the pear shape part of the ear. At the bottom of the 'antihelix' is a bump called the 'antitragus'. The more you familiarise yourself with the 'flow' of the ear, the easier it is to sculpt a successful replica.

55

When reconstructing multiple skulls, to save time, sculpt a couple different sized ears, make moulds for the different-sized ears, and cast them with melted clay.

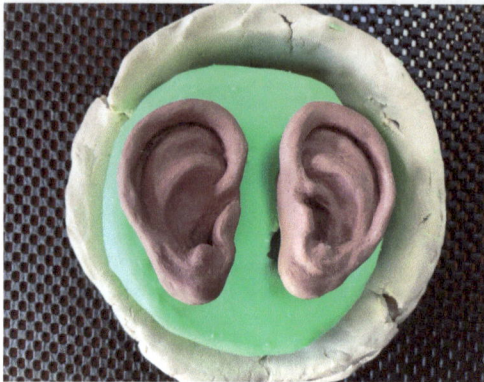

Ears cast using melted clay

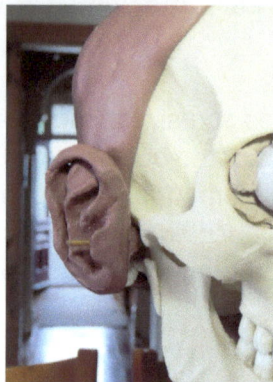

Ear secured in place

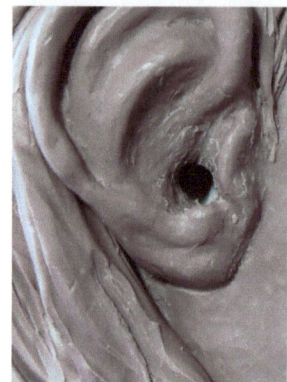

Auditory meatus

To determine the ear placement, run a thin wooden skewer through the auditory meatus (ear canal), from one side to the other, as a guide and then attach the ears to the skewer. This trick ensures the ears are in the correct position and the skull can still be seen through the ear hole.

* Adjustments and fine-tuning to the size and shape of the cast ears can better reflect each skull, such as the angle of the ear on the head or its overall proportions.

* In general, the ear size is determined by measuring the distance between the top of the eyes and the bottom of the nose.

* To angle the ear correctly, the front of the ear follows the same line as the jaw (mandible).

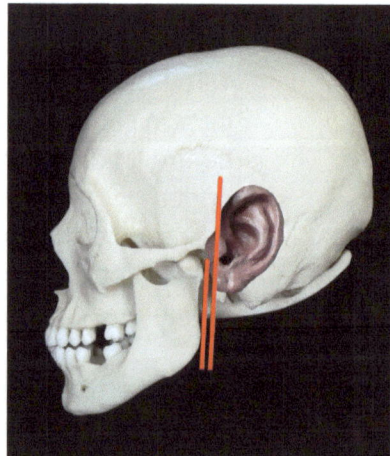

The front of the ear follows the angle of the mandible

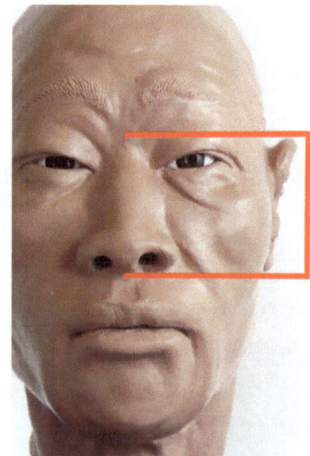

Approximate ear length

56

Additionally, the angle at which the ears protrude from the head can vary, with some ears being more prominent than others. Are the ears flat to the head, or do they stick out? Is the earlobe attached, or is it unattached?

Here are a few tips to determine ear shape and position.

* A strong and protruding supramastoid crest on the temporal bone causes the upper part of the ear to stick out (protrude).

* A rough outer surface of the mastoid process causes the lower part of the ear to protrude.

* When both of these characteristics are present, the ear will fully protrude.

* If the mastoid process faces downward, the earlobe is attached, while the earlobe is unattached if it angles forward.

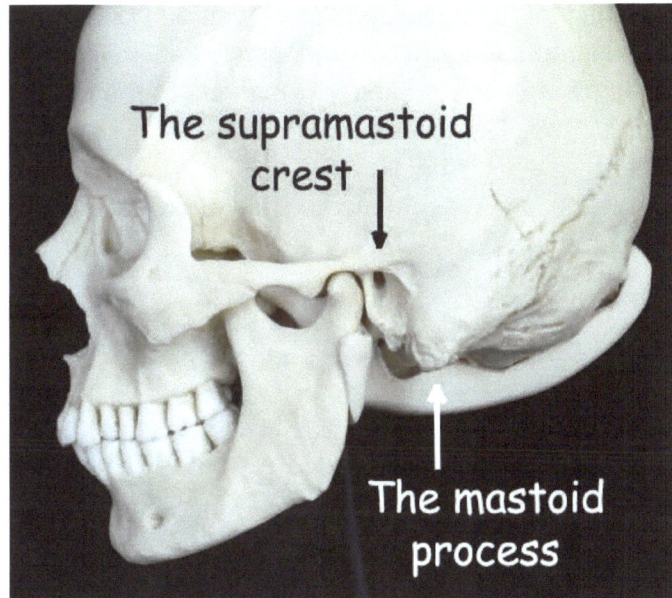

* Small mastoid processes that point inward towards the skull indicate small ears close to the head, while large and prominent mastoid processes suggest large and spread-out ears.

Tissue depth markers

Tissue depth markers, which correspond to the average tissue depth for the age, weight, and ancestry of the deceased, are placed on specific locations on the skull. You will find that the tissue depths reported in various studies are in millimetres. So you will need a good measuring tool to cut your tissued depth markers to the correct measurement. You will also need a sharp cutting tool, like a utility knife.

Once the tissue depth markers are cut to the correct length, various ways to attach them to the skull are possible.

If you are working with a plaster-cast skull, you can drill small holes at the tissue depth points and glue small wooden pegs into the holes.

Because I worked with finely cast resin replica skulls and did not want to damage them, I used glue gun sticks, cut to length, and secured them to the surface of the skull with sticky wax. This non-permanent fixture allows you to temporarily remove any tissue depth markers that are in the way while you are working. You can find a variety of tissue depth studies online, but most of them have come from the textbooks listed in the back of this book.

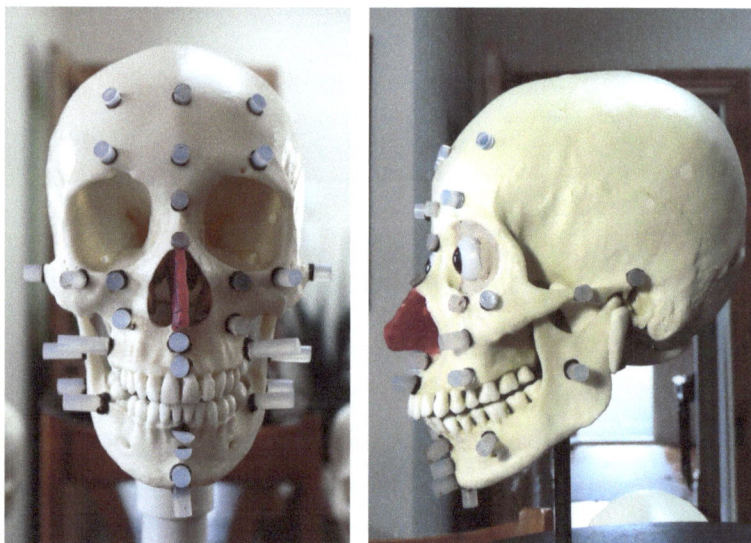

✳ All tissue depth measurements are in millimetres.

Tissue depth markers made from glue-gun sticks, attached to the skull with a sticky wax

58

Tissue depth marker locations by number

1	O	supraglabella	
2	G	glabella	
3	N	nasion	
4	RHI	end of nasal	
5	MP	mid-philtrum	
6	LS	upper lip margin	
7	LI	lower lip margin	
8	LM	labiomental - lip fold	
9	M	mental eminence -	
10	MN	(menton - beneath chin) - gnathion	
11	FE	lateral forehead	2
12	OS	(mid)-supraorbital	2
13	OR	orbitale - sub	2
14	ZA	zygomaxillare - inferior	2
15	LO	lateral orbit	2
16	ZY	lateral zygomatic - mid	2
17	SG	zygomatic - supraglenoid	2
18	GO	gonion	2
19	1M	supra M2 molar	2
20	MM	midmasseter - occlusal line	2
21	2M	sub M2 molar	2

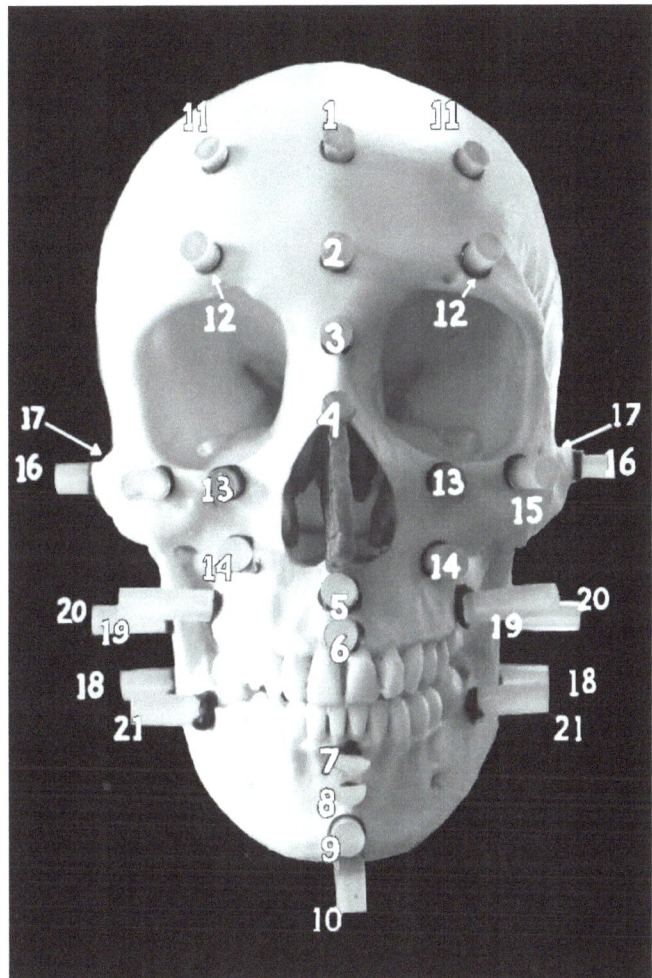

Tissue depth locations description

1	O	supraglabella	above the glabella - mid-forehead	
2	G	glabella	most prominent midline point between eyebrows	
3	N	nasion	midline point of nasal root	
4	RHI	end of nasal	end of nasal bone - cartilage/bone junction	
5	MP	mid-philtrum	maxilla midline, as high as possible before anterior nasal spine curve	
6	LS	upper lip margin	mid-upper vermilion line	
7	LI	lower lip margin	mid-lower vermilion line	
8	LM	labiomental - lip fold	mid-labiomental groove	
9	M	mental eminence - pogonion	anterior midpoint of chin, most projecting	
10	MN	(menton - beneath chin)	lowest medial landmark beneath chin	
11	FE	lateral forehead	projections on forehead above mid-eyebrows	
12	OS	(mid)-supraorbital	highest point on centre upper margin of orbit	
13	OR	orbitale - sub	lowest point on centre lower margin of orbit	
14	IM	zygomaxillare - inferior malar	below zygomatic on vertical line with infraorbital & supraorbital	
15	LO	lateral orbit	drop line down from outer orbit margin - place 10mm below orbit	
16	ZY	lateral zygomatic - mid	halfway on arch - most prominent point	
17	SG	zygomatic - supraglenoid	above & slightly forward of external auditory meatus - deepest point	
18	GO	gonion	most lateral point on mandibular angle	
19	1M	supra M2 molar	above 2nd maxillary molar	
20	MM	midmasseter - occlusal line	point on occlusal line - centre of ramus / mandible	
21	2M	sub M2 molar	below 2nd mandibular molar	

Nose

The nose is a complex organ of cartilage, fatty tissue, muscle and skin that protrudes from the nasal opening.

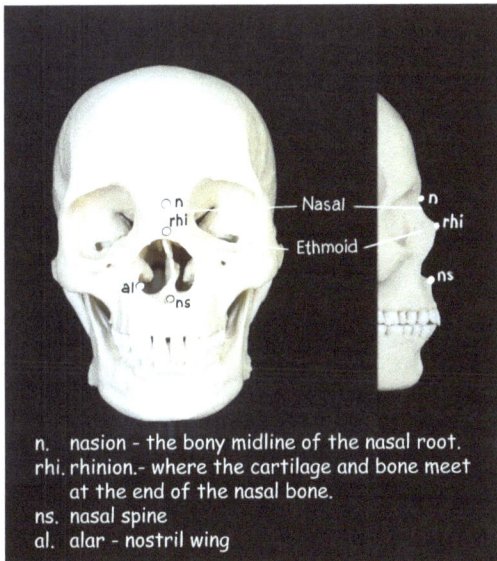

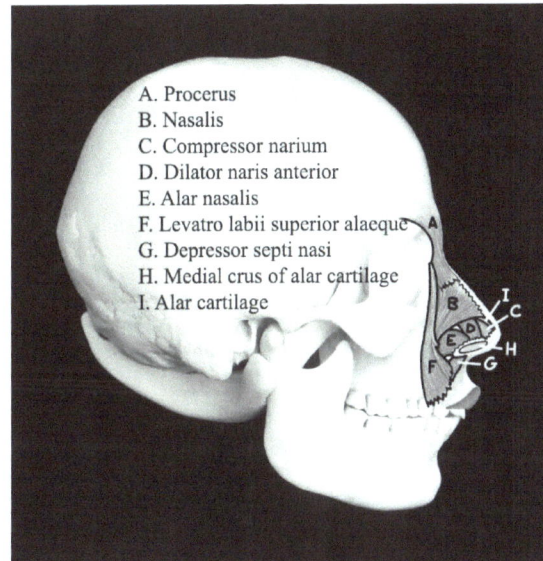

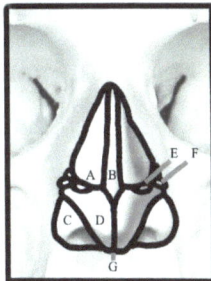

n.   nasion - the bony midline of the nasal root.
rhi. rhinion.- where the cartilage and bone meet
     at the end of the nasal bone.
ns. nasal spine
al. alar - nostril wing

A. Procerus
B. Nasalis
C. Compressor narium
D. Dilator naris anterior
E. Alar nasalis
F. Levatro labii superior alaeque
G. Depressor septi nasi
H. Medial crus of alar cartilage
I. Alar cartilage

Front View

A. Lateral nasal cartilage
B. Septal cartilage
C. Alar fibrofatty tissue
D. Greater alar cartilage
E. Accessory alar cartilage
F. Lesser alar cartilage
G. Anterior nasal spine

Inferior View

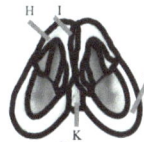

H. Lateral crus
I. Medial crus
J. Alar fibrofatty tissue
K. Septal cartilage -
(attaches to the Anterior
nasal maxilla of the skull)

The nose placement and shape is determined by several different formulas:

✳ The bridge of the nose can estimate the height, projection, length, and curvature.

✳ The angle of the nasal spine, can help determine whether the nose is turned up or turned down.

✳ The width of the nasal opening can estimate the overall width of the nose.

61

A projection, from the end of the bony part of the nose (top of the nasal aperture or opening) intersects with a line projected out from the angle of the nasal spine.

It is estimated that the distance from the tip of the nasal spine to the tip of the nose, when added to the average tissue depth marker, is approximately three times the length of the nasal spine projection.

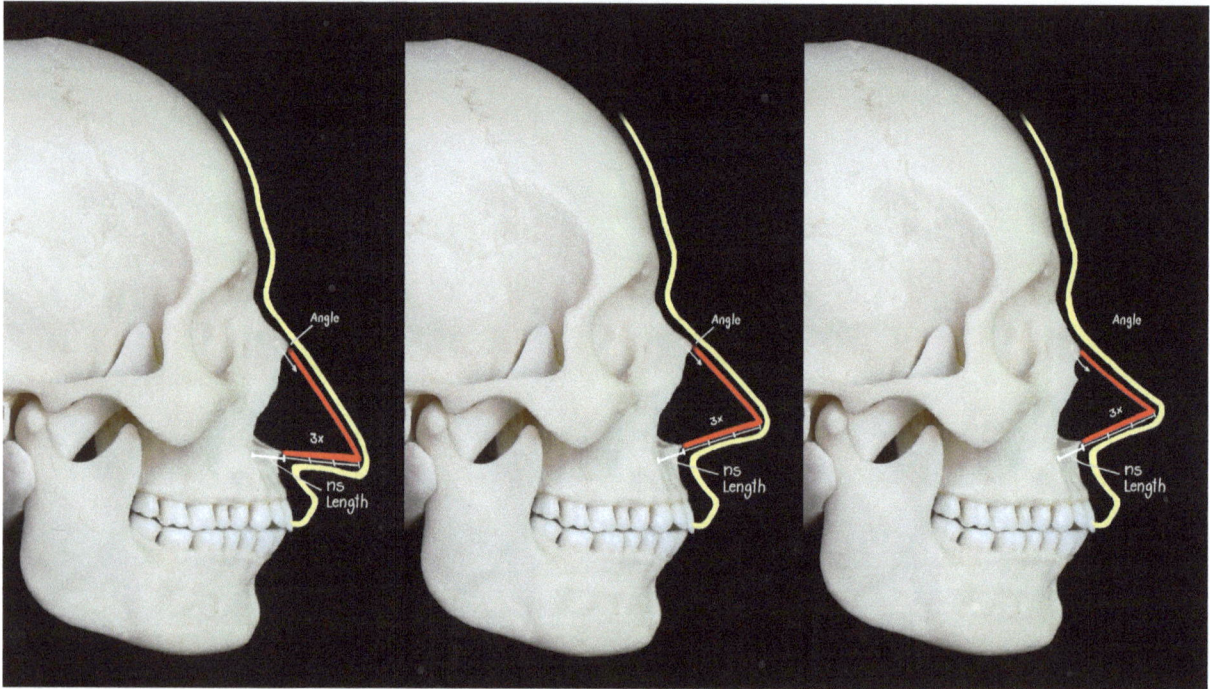

Calculating the the nose projection (3x length of nasal spine). Length and 'tilt' is determined by the intersection of the protected angle of the nasal spine (ns) and the angled line extended down from the rhinion at end of the nasal bone.

Contrary to other methods where the nose is fashioned from a solid block of clay, I created a flat 'septum' from a red wax, which followed the line and curvature of the nasal spine and the end of the rhinion (nasal bone). From there, I calculated the profile, the projection, and any deviation from the midline, if applicable (e.g. in a deviated septum or pronounced asymmetry in the nasal opening), using various methods as reference.

The cartilage in the nose plays a crucial role in shaping the overall structure and appearance of the nose. The nasal cartilage provides support and structure to the nose while allowing flexibility and movement.

To double-check the projection and profile I used the Lebedinskaya method[1] to predict the profile of the nose. This method involves drawing two lines. Line A runs from the nasion through the nasal spine to the, which is the front of the maxilla (the upper lip margin at the top of the front teeth). Line B runs parallel to line A, down from the foremost point on the rhinion.

Four to six equidistant lines are drawn perpendicular to lines A and B between the rhinion and the base of the piriform (nasal) aperture. At each perpendicular line, measure the distance from Line B to the piriform rim, and then add the same measurement to the other side of Line B, creating the outline of the profile.

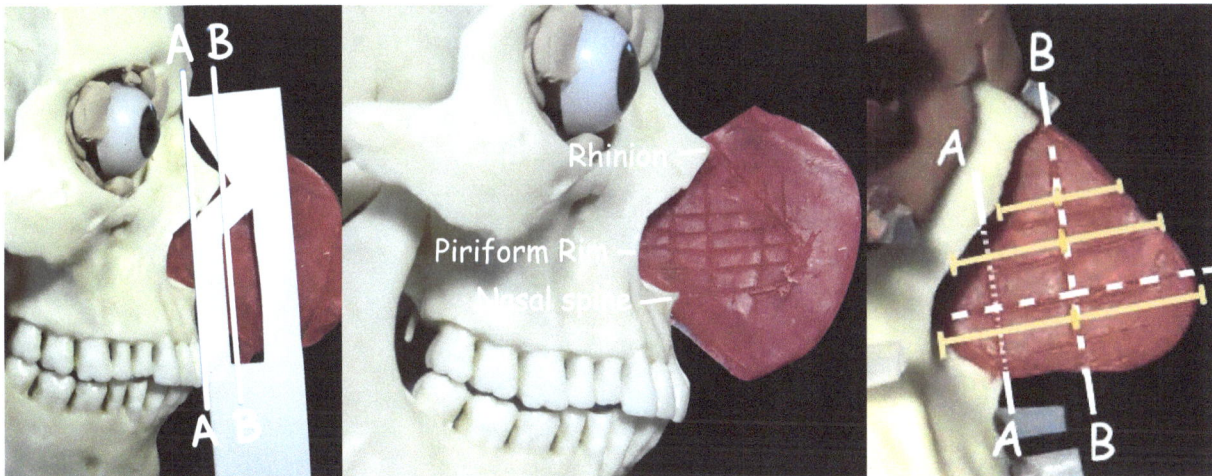

Line A - nasion to maxilla.   Measure the piriform rim to line B. Add the same measurement to the
Parallel line B from rhinion   other side of line B to create the nose's profile.

It was challenging applying the Lebedinskaya method measurements on a 3D skull - without the aid of a 2D image, so I fashioned a piece of card that would sit at the foremost point of the nasal bone and the upper lip margin, bypassing the projection of the end of nasion. This piece

Lebedinskaya G.V., Balueva T.S., Veselovskaya E.V. (1993) Principles of Facial Reconstruction. In: Işcan, Helmer (eds) *Forensic analysis of the Skull.* New York, Wiley-Liss, pp. 183–98. Google Scholar

of card allowed me to create line A on the wax septum and provided a cutout in which I could draw the parallel line B. I could then determine the perpendicular lines, take the measurements from line B and the piriform rim, add the same measurement to the other side of line B, and then trim off the excess wax, leaving the estimated nose profile.

Covering the nose aperture and the septum with a thin layer of clay, equivalent to the tissue depth at the end of the rhinion (rhi), and roughly forming the nostrils, works well. This method allows access to the nasal passage and the ability to reconfirm the nose's position, tissue depths, asymmetries, and nasal width as you proceed with the reconstruction.

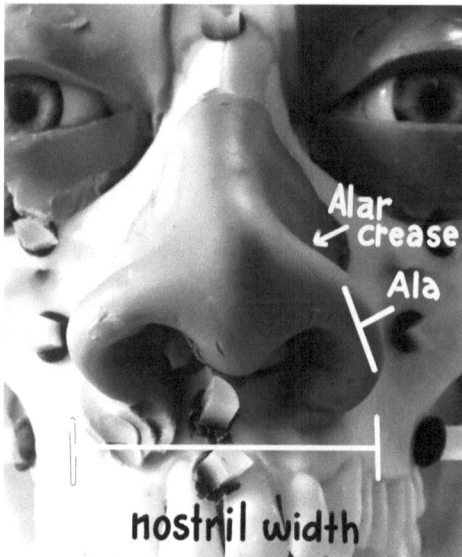

Calculate nostril width, measure nasal aperture ÷ 3, then x 5 = total width

✳ The nostril is called the 'ala', the crease at the top of the nostril is called the 'alar crease', and the pair of ala is 'alae'.

Typically, the widest point of the nasal aperture is three-fifths of the overall width of the nostrils.

To calculate the total width of the nose, measure the nasal aperture (opening) at its widest point (horizontally), divide the measurement by three, and multiply by five.

✳ The top of the nostril 'wings' can be found where the rim of the nasal cavity changes direction. This is the height of the nostril's wings at the alar crease.

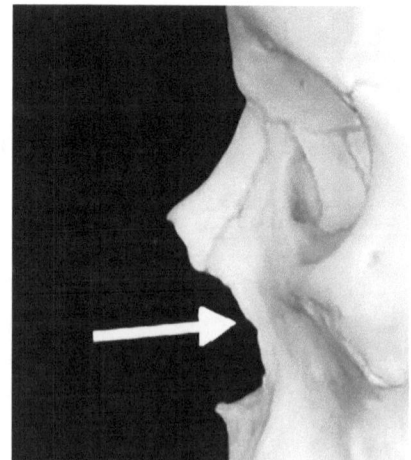

✳ Shape the alae (nostrils) by setting them in the natural groove that runs on the outside, and lower third of the nasal aperture, down to the bottom, and under the nasal opening to the nasal spine.

Subtle change of direction indicates height of the alae - nostril wings

64

Muscles of the face

The muscles of the face are not very prominent compared to other muscle groups in the body. This attribute is because the facial muscles are thin, flat, and primarily located near the skin's surface or covered with fatty tissue, like the buccal fat pad in the cheeks. Additionally, they are relatively small and do not have a large mass like the legs, arms, or back muscles.

Although the facial muscles are not prominent, they are responsible for many functions. Their subtle movements and contractions allow for the expressions of emotions and nonverbal communication. While the muscles of the face perform a range of functions, they are not typically very noticeable or prominent. However, in their totality, they do shape the face.

The following images depict some of the more prominent facial muscles that you might like to incorporate into your reconstruction.

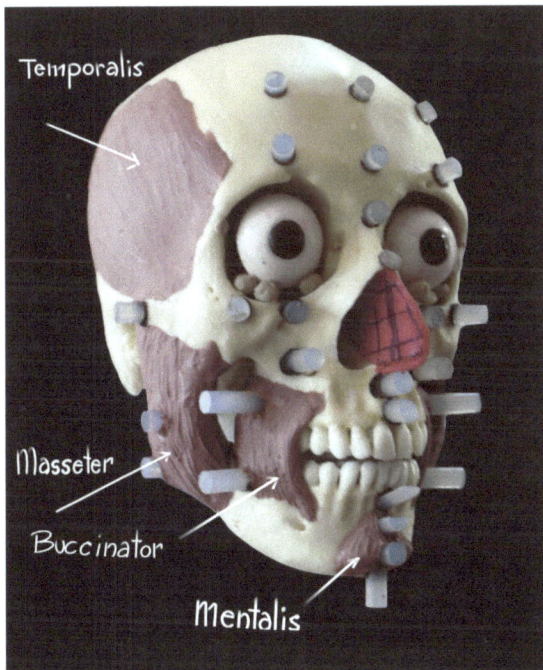

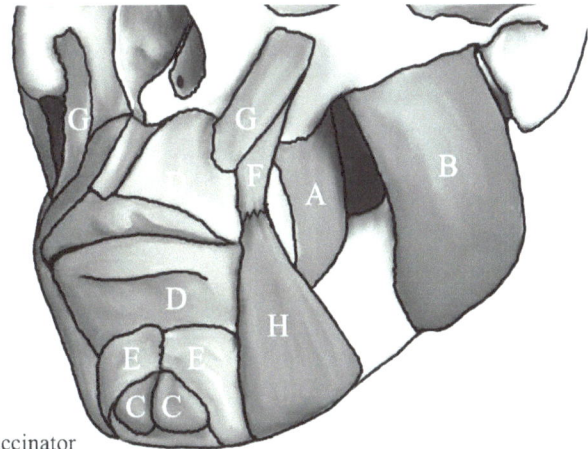

A-buccinator
B-Masseter
C-Mentalis
D-Orbiularis Oris
E-Depressor Labii Inferioris
F-Levator Anguli Oris
G-Levator Labii Superioris
H- Depressor Anguli Oris

The Muscles of the face

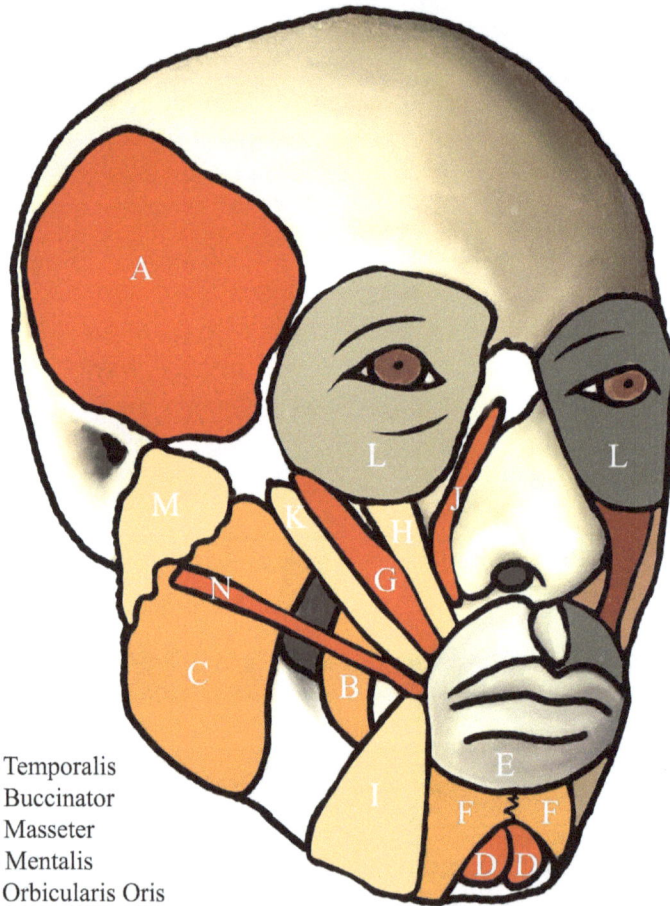

A - Temporalis
B - Buccinator
C - Masseter
D - Mentalis
E - Orbicularis Oris
F - Depressor Labii Inferioris
G - Levator Angulis Oris (not shown) sits under Zygomatic Minor (shown)
H - Levator Labii Superioris
I - Depressor Anguli Oris
J - Levator Labii Superioris Alaeque Nasi
K- Zygomatic Major
L - Orbicularis Oculi
M - Parotid Gland
N - Risorius

Joining the dots - Applying the clay

It is up to each artist to decide which reconstruction method works best. Each has advantages and disadvantages. However, both of these methods can be used in conjunction with each other to ensure that the anatomy is correct and streamline the sculpting process.

✳ The American Method is quicker, but it can leave room for more inaccuracy in the final result when applied by an artist less experienced in anatomy.

✳ The Manchester Method ensures the artist understands the anatomy well and accurately represents the underlying musculature, but much of this detailed work is then obscured with the addition of facial fat, such as the buccal fat pad in the cheeks.

Refer to diagrams or anatomy books for muscle placement, but always refer to the skull in front of you to locate the exact muscle attachments (origins and insertions) as discussed earlier. Always keep in mind the tissue depth markers as you build the muscle volume.

Working through this process, it became clear that it was optional to have the muscles precisely formed or finished, as additional tissue would be placed on top of them.

The muscles provided a suitable substructure and guide as to how the overlying tissue would behave, and I preferred the Manchester to the American method, which lays out the clay to the depth markers in the first instance.

✳ After constructing the musculature substructure, rough out the remaining fleshy areas with clay.

✳ Oil-based clay needs to be heated to become more malleable, so laying on large swathes of clay tends to be less precise and more roughcast.

✳ Any area with underlying muscle that doesn't come up to the height of the depth marker, such as in the cheeks, will be filled with fatty tissue and covered with skin.

Be mindful to keep the depth markers visible while you work. If you have removed them to work on other areas, check that you have the correct depths with either a small skewer or callipers.

✳   The image on the left is an example of the Manchester Method, where the underlying muscle is sculpted first.

✳   The image on the right is an example of the American method, where clay is put directly onto the skull, up to and around the depth markers.

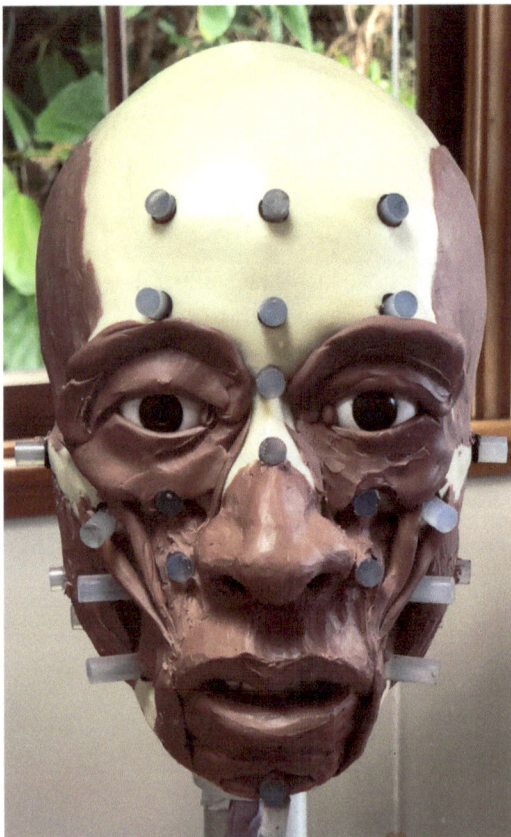

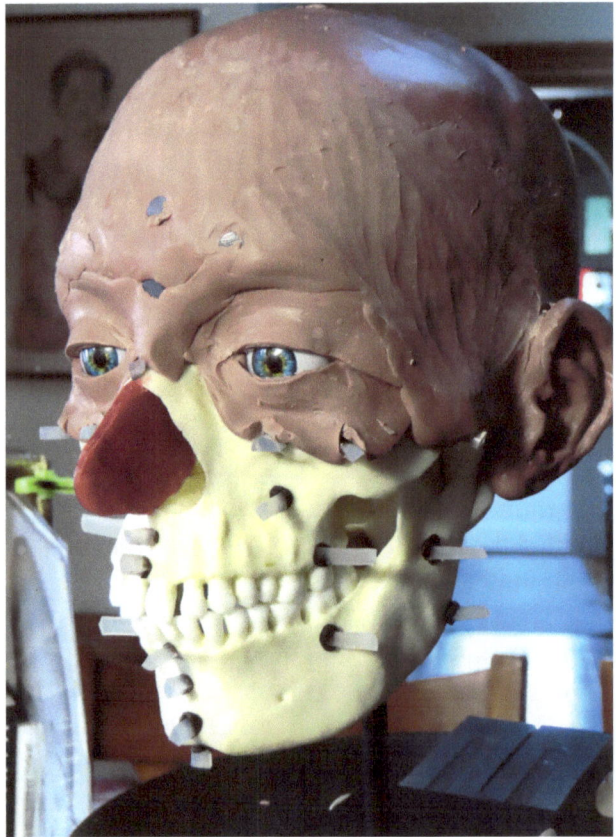

Manchester Method vs American Method

The mouth and lips

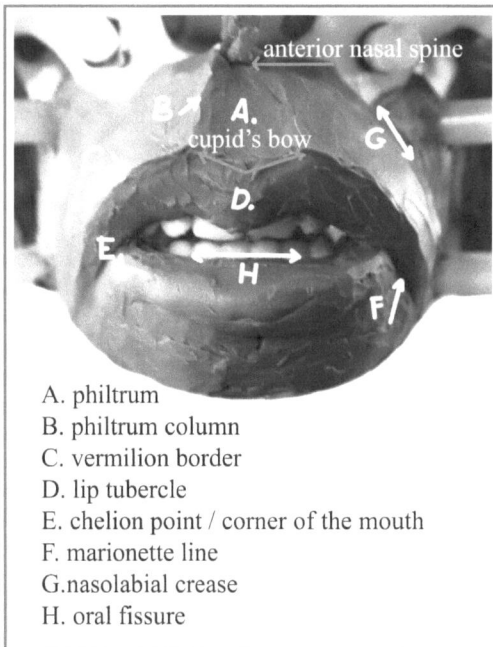

A. philtrum
B. philtrum column
C. vermilion border
D. lip tubercle
E. chelion point / corner of the mouth
F. marionette line
G. nasolabial crease
H. oral fissure

The mouth consists of many different parts. The 'orbicularis oris' is created by segmented quadrants of muscle that form a cohesive circular group that surrounds the mouth and covers the teeth (see muscles of the face). It has no attachment to any bone.

The lips are full and rounded, though they tend to thin with age. The lips are textured with vertical lines and the rim of the lips are lined with the 'vermilion border', a line denoting the distinction between the skin of the face and the skin of the mouth, particularly on the upper lip.

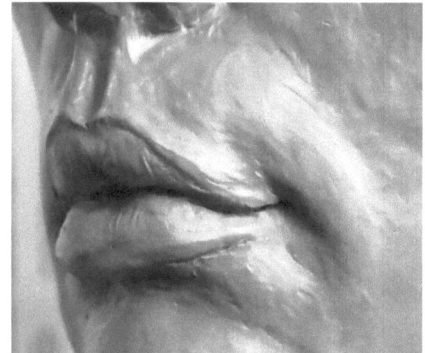

* The vermilion border is often more pronounced on those of African heritage and less prominent on those with European ancestry.

* In general, each lip is as wide, from top to bottom, as the corresponding teeth are long.

The lip thickness corresponds to the size and position of the teeth. Are the teeth large? If so, the lips are probably thicker. Do the teeth protrude (prognacy)? If so, the lips will also protrude.

The 'chelion point' is the corner of the mouth where muscles such as the 'risorius' meets (see muscles of the face) and are responsible for creating facial expressions, such as smiling. This point creates a complex area of fullness (amplitude) with curves that fold into one another.

The 'marionette lines' are created by the 'depressor angle oris' muscle that runs vertically from the corner of the mouth bottom to the bottom of the jaw. This crease tends to become more pronounced with age.

The 'nasolabial crease' is another area created by the meeting of the facial muscles responsible for facial expressions. With increased age, it can become a deep groove running from the side of the nose down the side of the face.

The mouth opening is called the 'oral fissure'. The width of the oral fissure corresponds to the juncture of the canine teeth and the first premolar (approx. the full width of the first six front teeth in the mouth).

The 'philtrum' is the groove that runs from under the centre of the nose (anterior nasal spine) to the top of the lip.

The philtrum is generally as wide as the two front teeth. The points at the bottom of the philtrum and above the mouth create the 'cupid's bow'.

When sculpting the mouth, place the bottom lip on first, as the top lip folds over the bottom lip.

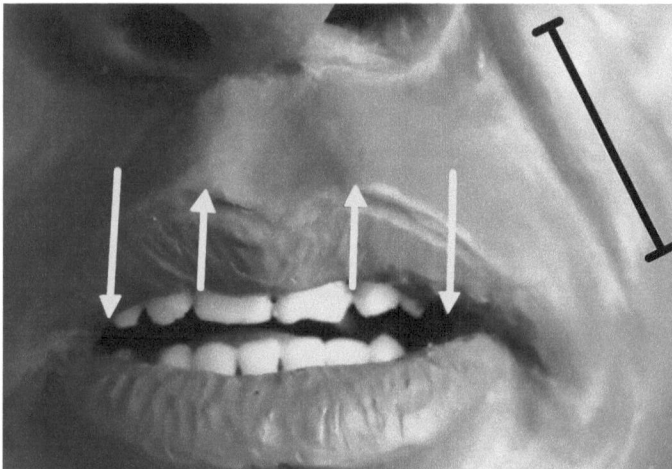

Philtrum and corner of mouth widths. Nasolabial crease from nose to corner of the mouth

Top lip over bottom lip and 'marionette lines' from corner of mouth down to jaw

Choosing an expression

Consider using a more neutral facial expression when reconstructing a skull. A neutral expression ensures that the facial reconstruction represents the most probable appearance of the individual without any bias towards a particular expression or emotion.

Some forensic facial reconstructions depict the deceased smiling, which can be unsettling. However, it is sometimes appropriate to show the teeth, particularly if they have unusual characteristics that might lead to an identification. In that instance, it is appropriate to sculpt the mouth partially open.

Maintaining a neutral expression while sculpting a forensic facial reconstruction is more appropriate for several reasons, including the accuracy of the reconstruction and ethical considerations.

One of the primary reasons for maintaining a neutral expression when sculpting a forensic facial reconstruction is to ensure accuracy. Given that the tissue depth studies in this process are most often derived from deceased individuals, it is best to represent the facial reconstruction with a neutral expression to preserve accuracy and consistency with the underlying anatomical data from these studies.

Maintain a neutral expression. Mouth partially open shows teeth.

The human face is incredibly complex, and even subtle changes in the positioning of the muscles can dramatically alter an individual's expression. Therefore, if the sculptor imposes a facial expression on the reconstruction, it could misrepresent tissue depth data, affecting the victim's appearance and leading to an inaccurate identification.

Additionally, a neutral expression may help reduce the reconstruction's emotional impact on the victim's family members, who may find it challenging to view an image of their loved one with a specific emotional expression.

Forensic artists must be objective and impartial when producing their work. By maintaining a neutral expression when sculpting a facial reconstruction, the artist avoids any biases from imposing an expression on the deceased's face. A more neutral facial expression is more likely to produce a faithful representation of the individual's appearance.

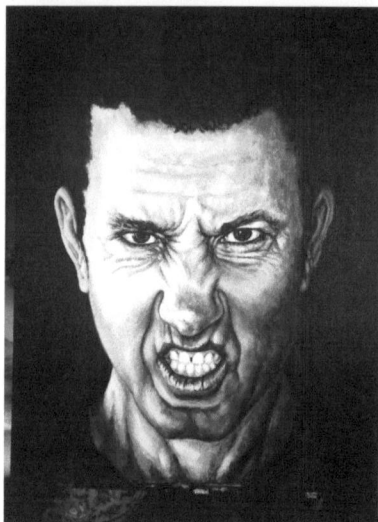

Facial expressions can significantly alter the overall appearance of the face, which may lead to inaccurate identification. Unless all facial muscles correspond to a specific expression, imposing a facial expression can make the reconstruction look odd. Sculptors must consider the subtle nuances of each muscle in the face and how they work together to create a natural appearance. Any deviation from this can make the reconstruction appear unnatural.

Imposing an expression on a reconstruction can mask accuracy if only some of the muscles involved in creating that expression are engaged.

Maintaining a neutral expression when sculpting a forensic facial reconstruction is recommended for accuracy, ethics, and ensuring a natural-looking result. Sculptors must remain objective and impartial while working, considering each muscle's placement and the nuances of the face to ensure that the reconstruction remains as faithful a representation of the deceased's appearance as possible.

Teeth

While this book does not delve deeply into the topic of teeth, it is still valuable to have a basic understanding of which tooth is which, as they serve as reference points within the mouth and play a significant role in shaping the lips.

Odontology is a specialist field unto itself. However, I encourage you to familiarise yourself and examine the teeth in the skull, noticing the wear patterns, missing teeth, etc. This observation will give you clues about the age at death and overall health.

Odontology, also known as forensic dentistry, studies teeth and their role in identifying human remains. The field of odontology has a long and fascinating history, with the use of dental evidence for identification purposes dating back to ancient times.

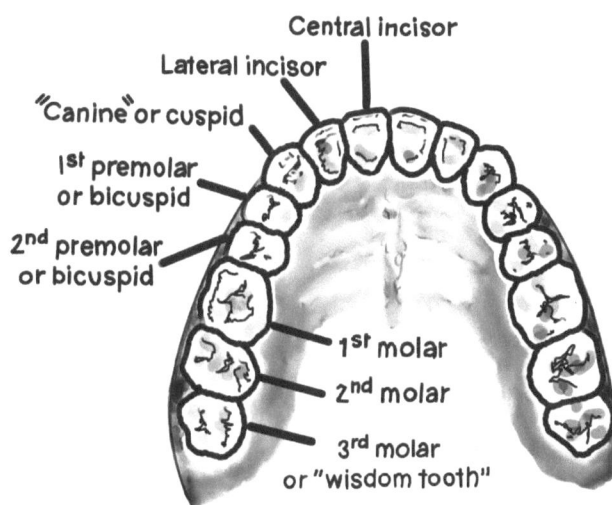

The Greek philosopher Hippocrates was among the first to recognise the unique characteristics of teeth, while the Roman physician Galen used dental records to identify bodies. In modern times, odontology gained significant prominence during World War II, when forensic dentists were used to identify the remains of soldiers. Today, odontology is critical in identifying victims of disasters, mass casualties, and criminal cases.

By analysing dental records, odontologists can provide valuable information that can aid law enforcement and bring closure to victims' families. The field of odontology continues to evolve, with new techniques and technologies constantly being developed to improve the accuracy and reliability of dental identification.

The eyes

The 'orbicularis oculi' muscles surround the eyes. This thin layer of muscle works similarly to the mouth muscles, being a muscle that surrounds an opening and works to open and close the orifice in which the eyeball sits.

The lateral canthus (side) and medial canthus (middle) is the corner where the top and bottom lids meet. In profile, the lateral canthus is further back than the medial canthus, which is closer to the frontal plane of the face.

The medial and lacrimal caruncles are located at the lacrimal crest (lc). The lacrimal caruncle is the small ball of pink tissue in the inner corner of the eye that protects the glands beneath it.

The depression along the junction of the lacrimal and maxilla bone (the groove seen behind point 'lc') houses glands in the lacrimal sac that secrete fluids (tears and oil) that keep the eye from drying out. Tears formed there drain through a duct into the nasal cavity. This configuration is physiologically essential for the outer corner of the eye to be slanted higher at the malar tubercle (mt) so that fluids in the eye flow toward the inner corner of the eye.

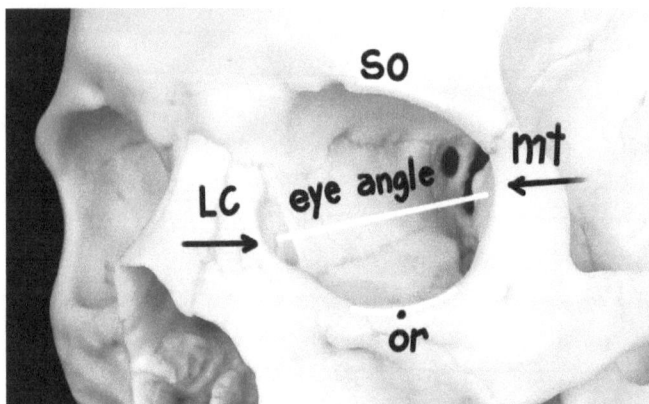

Lacrimal crest (lc). Malar tubercle (mt). Supraorbital rim (so). Orbital rim - lower (or). Eye lids angles down from mt to lc.

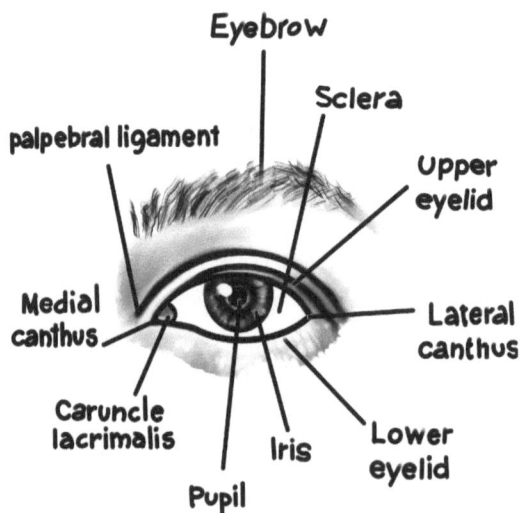

74

Some individuals look as if their eyes turn down on the outer corners. This illusion is due to how the eyelids sit rather than the lateral canthus being lower than the medial canthus.

＊ Place the bottom lid on your work before the upper lid.

＊ The lids meet at the medial canthus slightly lower and forward of the lateral canthus.

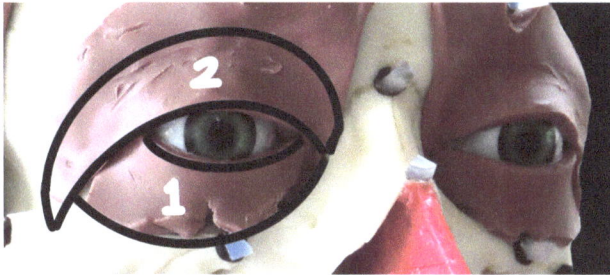

1. Place an arc of clay on the lower orbit
2. Place an arc of clay on the upper orbit
   overlapping the lower arc of clay

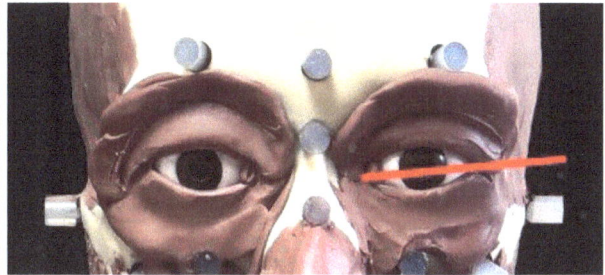

The lids meet at the medial canthus slightly lower
and forward of the lateral canthus.

The eyelid and eyebrows

The eyelid's position on the eyeball is a critical factor in conveying emotion through facial expressions. How the eyelid sits on the eye can significantly alter the appearance of the eye, which is a crucial element in expressing various emotions such as happiness, sadness, anger, surprise, and fear. For example, when the eyelids are lifted, the eyes appear wider, indicating surprise or excitement, whereas when the eyelids are lowered, the eyes appear smaller, suggesting sadness or anger. Therefore, understanding the importance of the eyelid's position on the eyeball is essential for accurately conveying the intended emotion through facial expressions, whether in social interactions, performances, or artistic expressions.

When reconstructing a skull, consider using a more neutral expression. To achieve a neutral facial expression, ensure that the lower eyelids sit just at the bottom of the iris and the upper eyelid sits midway between the pupil of the eye and the upper rim of the iris.

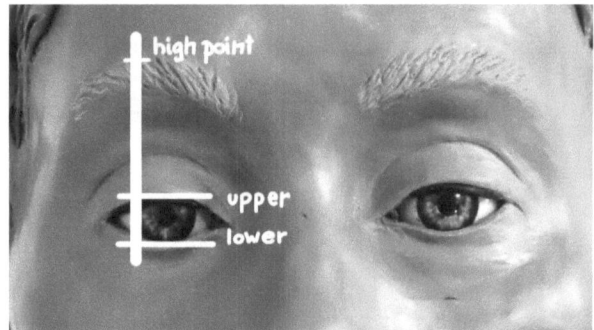

The highest point of the eyebrow is above the outside of the iris is inline with the highest point of the eyebrows.

The supraorbital ridge, the bony ridge above the eye socket is influenced by many factors, including genetics, age, and overall facial structure. This ridge also primarily determines the shape of the eyelid.

The supraorbital ridge is essential in shaping the eye area by providing a structural framework for the surrounding soft tissues. The shape of the ridge itself can vary significantly between individuals and can impact the overall appearance of the eye area.

For example, a prominent supraorbital ridge may create a more prominent brow bone and a deeper set eye, while a less prominent ridge may create a more rounded, softer eye shape.

The shape of the eye socket also determines the pattern of the folds in the eyelids and the way the eyelashes grow. For example, the lacrimal fold is a small ridge of skin that runs from the inner corner of the eye to the nose, giving it its unique shape and character. For those with Asian ancestry, this fold tends to be more prominent than in other ethnicities. It is also commonly called the 'epicanthal', 'epicanthic', or 'Mongolian' fold. The presence of the lacrimal fold can create the illusion of a smaller and more strabismic eye.

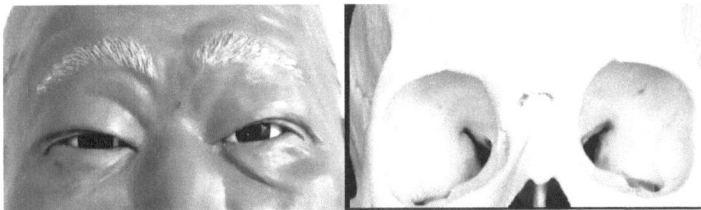

Ultimately, the eye socket is a critical factor in determining the shape and pattern of the eyelids, which gives each person their unique eye appearance.

...

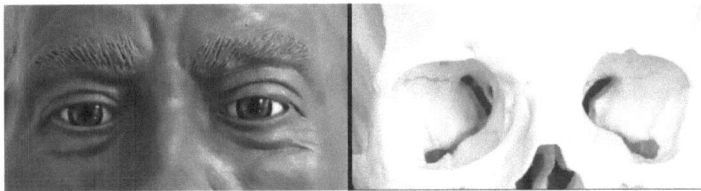

Eyebrows are a significant facial feature in terms of function and aesthetic appearance. They comprise dense, short hairs located above the eyes, on and just above the bony supraorbital ridge. The eyebrows follow the contour of the supraorbital ridge, and the underlying bony structure influences their shape and thickness.

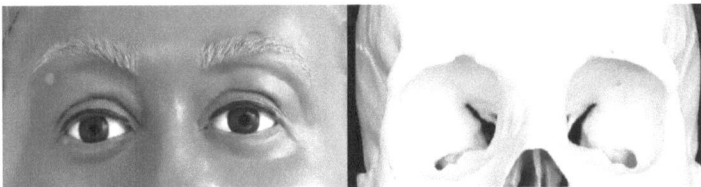

Examples of eyelid and eyebrow patterns compared to their respective skull (top - Asian male, middle - European male, bottom - African/American female).

The primary function of eyebrows is to protect the eyes from sweat, dirt, and other debris that can fall from the forehead.

Eyebrows also play a crucial role in non-verbal communication, expressing emotions, and conveying mood. For instance, raising eyebrows can indicate surprise, while furrowing them can signify anger or concern. Eyebrows are also a significant component of facial symmetry and beauty, and their shape and thickness can significantly impact the overall appearance of a person's face.

Generally, the brow ridge is more prominent in males than in females, which can affect the shape and thickness of the eyebrows.

The shape of the brow ridge can also vary among individuals, which can affect the eyebrows' patterning. For example, some people may have a more pronounced arch in their brow ridge, resulting in a more dramatic arch in the eyebrows. Others may have a flatter or more sloping brow ridge, which can lead to a straighter or less defined eyebrow shape.

The patterning of the eyebrows follows the contour of the brow ridge and the supraorbital ridge. The underlying bony structures influence the shape and thickness of the eyebrows and can vary among individuals based on factors such as gender and the shape of the brow ridge.

Below are some combinations of attributes that make for various eyebrow patterns.

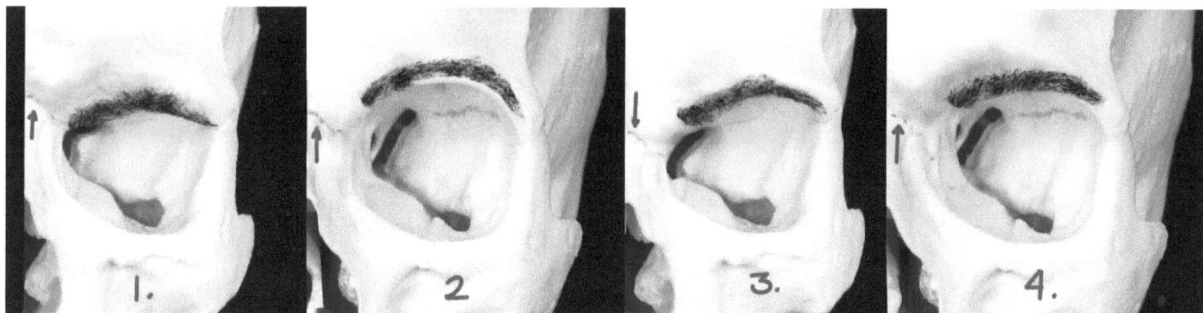

1.   A prominent brow ridge, high nasal root, and rounded supraorbital rim make for an eyebrow that tucks under the brow ridge and is lower overall.

2.   A smooth brow ridge, high nasal root, and sharp supraorbital rim make for a more rounded eyebrow.

3.   An average brow ridge, low nasal root, and sharp supraorbital rim make for a more arched eyebrow.

4.   A prominent brow ridge, high nasal root, and sharp supraorbital rim make for a straighter eyebrow lifted at its end.

The nose in detail

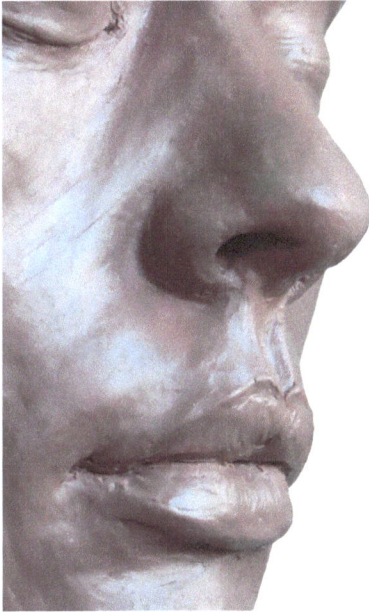

By now you will have a rough outline of the nose structure and the underlying bone and cartilage. The next step is to refine it.

Begin by sculpting the overall shape of the nose by applying clay by hand or with a sculpting tool. The nose should blend seamlessly into the surrounding facial features.

Check your work regularly and assess the progress from different angles to ensure that the proportions and symmetry (or asymmetries) are accurate.

Next, work on adding details to the nose. Refer to the diagrams of the nose in previous chapters. Add subtle curves and ridges (amplitude) to create a more lifelike appearance.

The size and shape of the nostrils, as well as the placement and angle of the tip should already be set out in the setting up of the nose.

The shape and size of the cartilage determine the height and width of the nose, as well as the curvature of the nasal bridge and the shape of the nasal tip. Additionally, the cartilage's thickness and density can influence the nose's overall aesthetic, including whether it appears more refined or bulbous.

The soft tissue at the end of the nose will correspond to the underlying bony nose. For example, if the nasal spine is bifurcated (split into two bony spines), it can indicate a cleft at the end of the nose.

A realistic nose should not be perfect - imperfections and irregularities add to the natural appearance.

Finishing touches

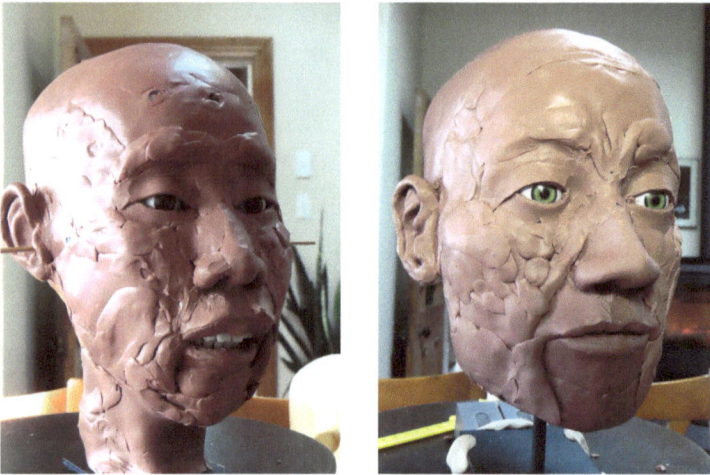

Fleshing out the skull by adding the fatty tissue

At this point, you will have filled out the face up to the tissue depth markers using the method of your choice.

Leaving the depth markers visible ensures that the clay is to the correct depth, but in the areas where there are no depth markers, continue to check the depth with the type of callipers that have a probe, or use any probe where you can accurately measure the depth of the clay.

Once all the components of the facial features are in place, and you are happy with the overall proportions, refining the elements, such as smoothing the skin or adding texture, wrinkles, and hair, can be the most time-consuming.

Texture can add depth and realism to the sculpture. Use a sculpting tool, sandpaper, a pot scrubber, or a brush to add texture to the skin, such as pores or wrinkles.

As we age, our faces undergo several changes that reflect the aging process, environmental factors, and lifestyle choices. These changes are visible as lines, furrows, grooves, and wrinkles that form on the face over time.

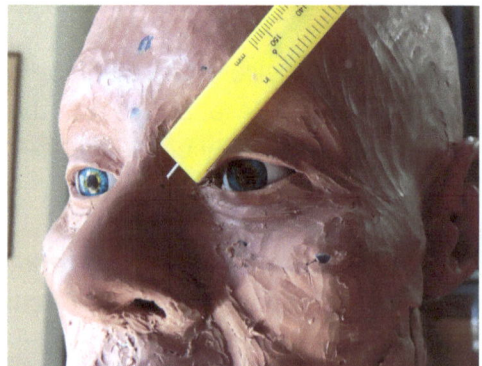

Checking tissue depths and leaving the markers visible

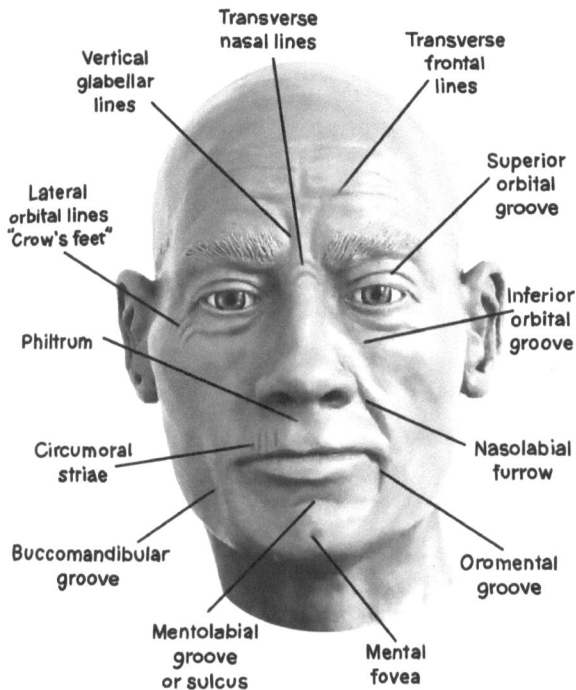

Transverse nasal lines
Vertical glabellar lines
Transverse frontal lines
Superior orbital groove
Lateral orbital lines "Crow's feet"
Inferior orbital groove
Philtrum
Circumoral striae
Nasolabial furrow
Buccomandibular groove
Oromental groove
Mentolabial groove or sulcus
Mental fovea

The reconstruction artist can better represent their subject by understanding and applying the details of how the human face ages.

Considering the age of the skull you are working on will direct you to the appropriate application of lines, wrinkles, and furrows.

However, some of these superficial soft tissue lines and grooves may be present at a younger age, particularly in males, because of heavy skull features, such as a heavy brow that forms transverse nasal lines and a prominent chin that accentuates the mentolabial groove. For example, a mental fovea, or 'dimple', may be present in the middle of the chin at a younger age.

Age 20-30 years old:
At this stage of life, the skin is generally firm, supple, and smooth. The face has a youthful appearance, with few visible lines or wrinkles. However, some fine lines around the eyes, such as 'crow's feet' and the nasolabial folds (the lines that run from the sides of the nose to the corners of the mouth), may begin to appear. These lines are usually subtle and not very noticeable.

Age 30-40 years old:
In the 30s, the skin begins to lose some elasticity, and the first signs of aging become more apparent. Fine lines and wrinkles may become more pronounced around the eyes and mouth. These lines are often referred to as "laugh lines" or "smile lines" and are a natural result of facial expressions. The nasolabial folds may deepen, and other lines may start to appear, such as the marionette lines (the lines that run from the corners of the mouth to the chin) and the vertical lines between the eyebrows, known as vertical glabellar lines.

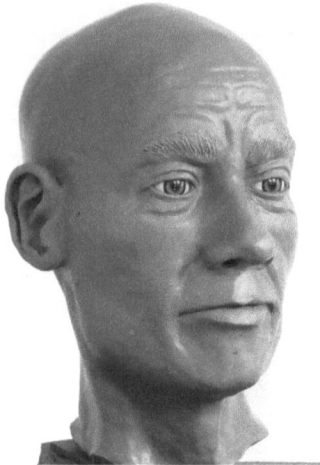

Age 40-50 years old:

In the 40s, the skin continues to lose elasticity, and wrinkles become more pronounced. The cheeks may begin to sag, and jowls may form along the jawline. The nasolabial folds and marionette lines may become even more noticeable. Additional lines, such as the horizontal lines on the forehead, may also appear, called transverse frontal lines or 'worry lines'. Circumoral striae lines, small vertical lines on the lips, may start forming around the mouth. These lines are often present in smokers.

Age 50-60 years old:

In the 50s, the skin becomes thinner and less plump, which can make wrinkles and lines even more apparent. The skin may also become drier and more prone to discolouration, age spots, and uneven tone. The nasolabial furrows and oromental grooves (marionette lines) may deepen around the mouth and jawline, and the jowls may become more prominent. The skin around the eyes may also start to sag, creating bags and dark circles, increasing the inferior and superior orbital grooves.

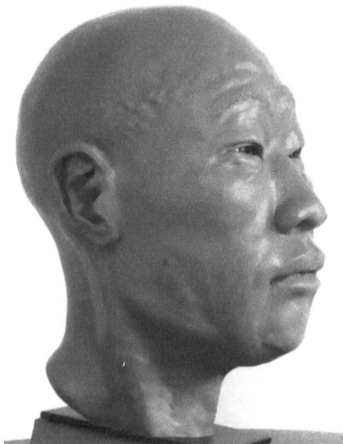

Age 60-70 years old:

In the 60s, the skin continues to thin, and wrinkles become more pronounced. The skin may also become looser, and the cheeks may appear sunken due to fat loss in the cheeks, increasing the circumoral striae groove. The jowls may become more prominent, and the neck may start to sag, forming what is commonly called a "turkey neck." The forehead wrinkles and frown lines may become more profound and more noticeable.

Age 70-80 years old:

In the 70s and beyond, the skin continues to thin, and wrinkles become even more pronounced. The skin may also become translucent, making the veins and bones beneath more visible. The skin around the eyes may become even looser, creating more pronounced bags and dark circles. The jowls and neck may become even more saggy and loose, and the chin may start to recede.

The Neck

A sculpted head without a neck always looks a bit macabre. So it is important to incorporate the neck into your reconstruction work. Not all neck muscles will be visible in your work, so only the most prominent muscles have been included in the following illustrations.

The neck is not just a cylinder of muscle around the spine. The neck muscles are a complex group of muscles that give the neck unique characteristics, provide stability, support, and movement to the head and neck vertebrae. This amazing combination of muscles is responsible for controlling the head and neck position, allowing us to move our head a great range of motion in various directions and provide stability during movements such as running, jumping, and bending.

If you choose to have your reconstruction looking to the side, different muscles will engage and become more prominent in your reconstruction than if it is looking straight ahead. Several of the visible muscles in the neck region include the trapezius, sternocleidomastoid, scalene, levator scapulae, and the muscles near the throat, such as the sternohyoid.

The trapezius is a visibly large, triangular-shaped muscle extending from the skull's base to the middle of the back and shoulders. It is responsible for moving and stabilising the shoulder blades and neck, as well as supporting the weight of the arms. One of its unique features is its attachment to the occipital bone, located at the back of the skull. Specifically, the upper fibres of the trapezius connect to the superior nuchal line of the occipital bone, a bony ridge that runs horizontally across the base of the skull, and its attachment point is clearly visible (chapter

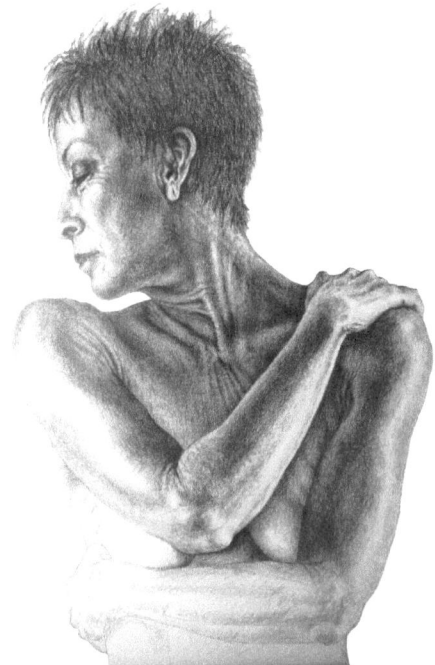

about muscle attachments). This connection allows the trapezius to assist in tilting the head back and maintaining an upright posture.

The sternocleidomastoid muscle is one of the largest muscles in the neck, located on each side of the neck. Its insertion runs from the base of the skull at the mastic process (behind and below the ear) to its origin at the sternum and clavicle bones. This muscle is responsible for tilting the head to the side and turning the head, and one that is very visible when engaged.

The scalene muscles are a group of three muscles located on each side of the neck. They run from the cervical vertebrae to the first two ribs. These muscles are responsible for elevating the ribs during inhalation, and they

The most visible muscles are the trapezius and the sternocleidomastoid. The sternocleidomastoid connects from the sternum and clavicle to the mastoid process.

also help to turn and bend the neck. While not as visible as the sternocleidomastoid, they fill the area between the sternocleidomastoid and the trapezius.

The levator scapulae also sit beside the scalene muscles. It is a long, thin muscle located at the back and side of the neck. Its primary function is to elevate the scapula or shoulder blade. The levator scapulae originate from the transverse processes of the upper cervical vertebrae and insert onto the superior angle of the scapula. It works with other muscles, such as the trapezius and rhomboids, to allow for scapula movements. The levator scapulae sit near the scalene muscles and, like the scalenes, also fill out the neck's volume between the sternocleidomastoid and the trapezius.

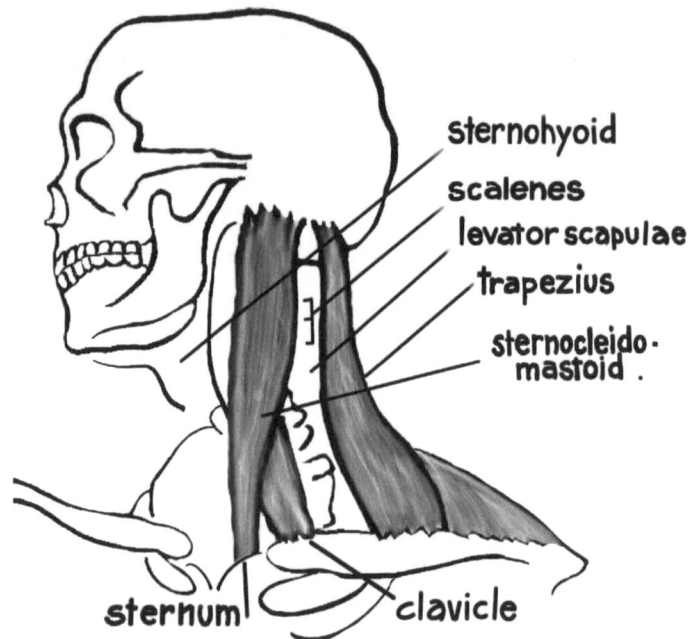

How long should you make the neck?

As a general rule, the length of an adult human neck is roughly equal to the length of the head. The average length of an adult human head is about 9 inches (23 cm) from the chin to the top of the skull. The length of the neck is also about 9 inches (23 cm) on average. Therefore, the ratio between the human head and neck length is roughly 1:1.

How big around should you make the neck?

The circumference of the average adult human head is roughly 21 to 23 inches (53-58 cm), while the circumference of the neck is usually around 14 to 16 inches (35-41 cm). Therefore, the ratio of the size of the human head to the neck circumference is typically around 1.5:1 to 1.6:1.

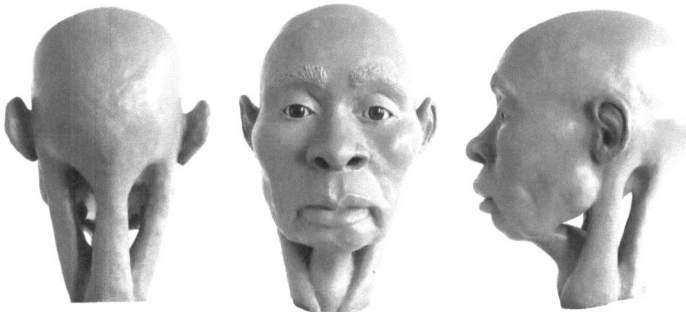

It's important to note that this ratio can vary based on factors such as posture, body type, and skeletal structure. Additionally, infants and children have proportionally shorter necks than adults, which gradually lengthens as they grow and develop.

Using a bit of artistic licence can suggest significant neck muscles and the overall shape of the neck.

Adding a neck to a head sculpture is visually essential as it adds structural stability and conveys a sense of completeness and realism. Without a neck, a head appears detached and unfinished, lacking the necessary anatomical features to make it look lifelike. Adding a neck provides a connection between the head and the body, allowing the sculpture to appear as a whole rather than a disconnected part.

Artistically speaking, the neck offers an opportunity to convey emotion and movement, with slight changes in angle or position creating a sense of tension or relaxation, and is crucial to creating a visually compelling and realistic piece of art.

Shoulders

Adding a suggestion of shoulders below the neck of a bust sculpture can significantly elevate its overall aesthetics.

Even a subtle "flair" at the bottom of the neck that implies the beginning of the shoulders, or lowering the neck enough to link the sternocleidomastoids to the clavicles, can enhance the piece's impact and authenticity.

This extra attention to detail imbues the sculpture with depth and dimensionality, breathing life into it and rendering it more true-to-life. As such, incorporating the suggestion of shoulders can make a great difference in the final appearance of the artwork.

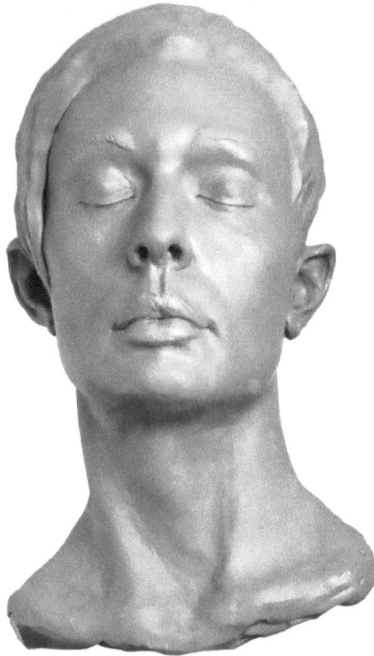

The suggestion of shoulders helps to balance the overall look of your work.

Hair

One of the most challenging aspects of facial approximation is recreating the person's hair, especially when no photographic history or forensic evidence is available.

Some reconstruction artists use wigs to represent the subject's hair. This method is fast and effective, but wigs are only sometimes available to amateur artists. In such cases, clay can be a more versatile and practical medium for creating the hair of the individual.

Here are some creative ways to create hair with clay for forensic facial reconstructions, keeping in mind the individual's ancestry:

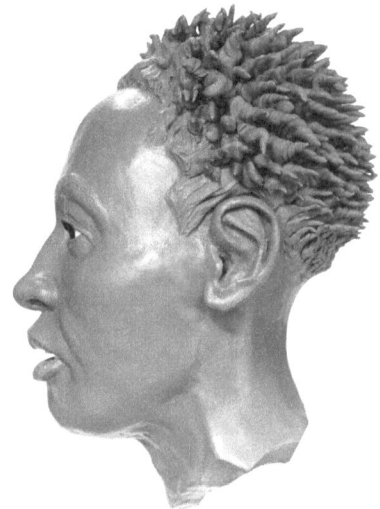

Oil-based clay allows for creativity in recreating the hair.

✳ Research: Start by researching the person's ancestry and cultural background. This can give you an idea of what type of hair they might have had, such as straight, curly, wavy, or coiled.

✳ Texture: Texture is a fundamental aspect of hair and varies depending on the individual's ancestry. For example, people with African ancestry tend to have coarse, curly hair, while those with Asian ancestry tend to have straight, smooth hair. To recreate texture in clay, use tools such as toothpicks, combs, and wire brushes.

✳ Colour: Hair colour is another identifying aspect that can vary depending on the person's ancestry. Using various-sized pieces of clay, and leaving gaps, conveys depth and contrast. Keeping the hair the same colour as the clay is a good option. In the same way, a black-and-white photograph implies the colour without it being explicit. If the skull is being used for identification, let those undertaking the identification fill in the blanks.

✳ Style: Hairstyles can reveal clues to a person's identity, social status, and cultural background. When practical, research the person's era and culture to get an idea of what type of hairstyle they might have had. Use clay to create the hairstyle, then, if needed, use a hair dryer to soften and manipulate the clay into the desired shape.

✳ Accessories: Hair accessories such as headbands, hats, and can also provide clues about a person's identity and cultural background. If there is an indication that these accessories were present or used traditionally, you can use clay to create and attach them to the head.

✳ Layering: Hair is rarely flat and uniform, so layering is an important technique to create a realistic-looking hairstyle. Use varying layers of clay to build up the hair, and use sculpting tools or other objects, such as toothpicks and wire brushes, to create texture and volume.

Creating hair with clay for forensic facial projects requires creativity, research, and attention to detail. By considering the individual's ancestry and cultural background and using tools and techniques to create texture, colour, style, and accessories, you can create a realistic-looking hairstyle that can provide additional clues to the person's identity.

Using a pasta-maker to create hair

Hair created with pasta-maker. Traditional cultural accessories.

Refining your work

Sculpting tools are an essential part of creating lifelike sculptures from oil-based clay. You can create a sculpture with a smooth surface, with wrinkles, pores, and aging skin. Selecting the right tools allows you to create a more natural appearance. Texture can add depth and realism to your work. Various sculpting tools, and even household items like a brush, can add texture to the skin.

* Select the right tools:
  The right tools will depend on the area of the face you are working on. Generally, any sculpting tools, including metal, wood, and even old dental tools, will suffice to create the fine detail and finished smoothness you require for a believable reconstruction. Ensure your tools are clean to achieve the best results, as oil-based clay can be very sticky.

* Prepare the clay:
  Oil-based clay can be challenging, especially if it's hard or too soft. For finer details, warm it up in your hands or heat it with a hairdryer or in the crockpot. However, be careful not to overheat the clay to avoid it from melting.

* Work in sections:
  Smoothing out the entire piece can be overwhelming, so work on various sections. You will already have the features formed, including the soft tissue that comprises most of the head. Now you can move on to the finer details. This approach lets you focus on each section and achieve the desired smoothness and lifelike appearance.

* Use a smoothing tool:
  A smoothing tool, like a flat wooden tool, is excellent for smoothing out the surface of the clay. Use gentle pressure to avoid pressing too hard and creating unwanted dents or marks when the clay is soft, but you can use the same tool with more force to burnish the surface to make it smoother when the clay is cold.

✳ Refine the details:
    Use smaller tools to add finer details to your
    sculpture, such as wrinkles, hair, or textures.
    Take your time and work slowly to ensure you
    achieve the desired level of detail.

✳ Use fine grit sandpaper or pot scrubber:
    Once you've finished smoothing the clay with
    your tools, you can use sandpaper or a pot
    scrubber to achieve a more realistic finish.
    Oil-based clay can be very shiny. Gently run a
    fine-grit sandpaper or a pot scrubber over the
    larger areas of skin to create a more natural-
    looking surface, giving your sculpture a
    lifelike appearance.

Most importantly, remember to take your time, work
slowly, and enjoy the process.

Tool marks give the impression of aging
skin and wrinkles.

Forensic facial reconstruction projects

The following are examples of seven skulls that I have reconstructed. The information chronicles the reconstructed skull, observational notes, tissue depth data, and the reconstruction process of each skull.

Each skull purchased from ©Bone Clones (excluding the elongated skull) came with an Osteological Evaluation prepared by Evan Matshes, BSc, MD Consultant Osteologist.

However, I made observational notes and reconstructed each skull before reading the information provided by ©Bone Clones. During the reconstruction process, I added additional observations in **BOLD**. After finishing the reconstruction process, I then read the Osteological Evaluation report, adding any information I missed in RED to indicate supplemental information.

I used as much data from tissue depth studies as possible, averaging the numbers to come up with the most accurate tissue depths possible, taken from different studies in different ways (CT scan, deceased with embalming, deceased without embalming, etc.)

✳ All tissue depth measurements are in millimetres.

If you purchase a skull from ©Bone Clones, I encourage you to make your own observational notes before starting your reconstruction project. The observational process allows you to get to know the skull intensively.

I encourage you to find various tissue depth studies online, in other books, or through libraries.

You are welcome to copy the spreadsheet template of the observational notes and use any of the tissue depth studies listed.

Elongated Skull - Female (over 2000yrs old) - Paracas, Peru

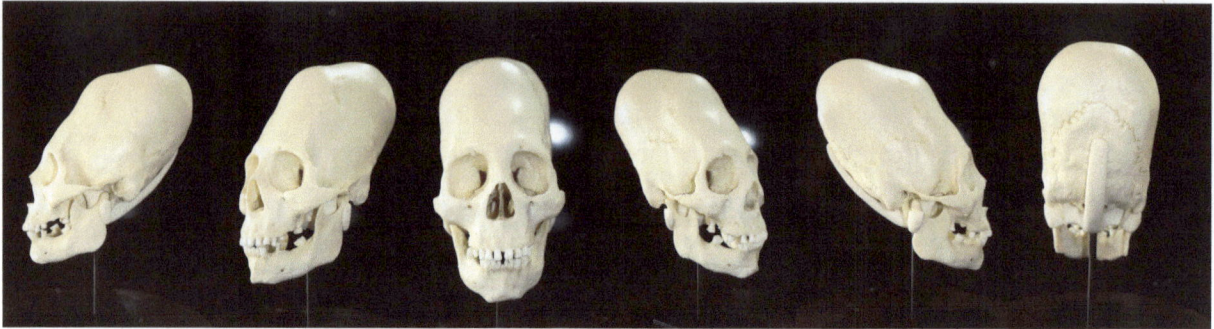

Elongated skull - replica supplied by ©Bone Clones

This reconstruction project is a female with an elongated skull from Paracas, Peru. This skull is reported to be over 2,000yrs old. Interestingly, I had actually seen and photographed this very skull (the original) on a trip to Peru in 2015. To find a replica of it was exciting.

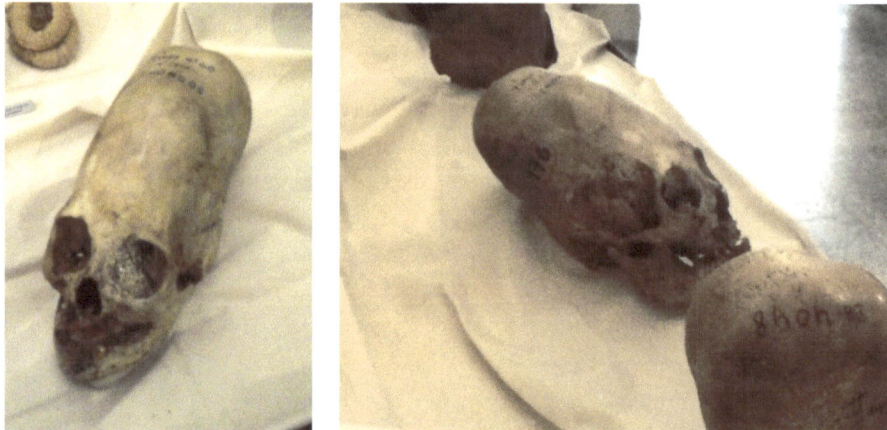

Original 2,000yr old female elongated skull - Paracas, Peru
Photoed by the author in 2015

Determining potential features by DNA

Elongated skulls, also known as cranial deformation, have been found in various parts of the world, including South America, Africa, and Europe. Elongating the head was achieved by tightly binding an infant's skull with a cloth or other materials, often from birth. The process gave the crown of the skull a particular shape, considered a symbol of social status or cultural identity.

One of the most famous examples of elongated skulls comes from the Paracas culture in Peru. The Paracas people were a pre-Columbian civilisation that inhabited the region from around 800 BCE to 100 BCE. Their elongated skulls are distinct from other cultures due to their size and shape, with some measuring up to 25% larger than a typical human skull.

The Peruvian archaeologist Julio Tello discovered the Paracas' skulls in the early 20th century. He found hundreds of elongated skulls in the Paracas Peninsula and surrounding areas. Since then, numerous studies have been conducted on these skulls to understand their significance and the methods used to create them.

One theory suggests that the Paracas people intentionally elongated their skulls to differentiate themselves from other cultures and to show their social status. Another theory suggests that the elongated skulls result from a genetic mutation or a natural deformation caused by head-binding practices that were prevalent in the region..

DNA testing of elongated skulls from Paracas, Peru have had surprising results. The test results suggest that the maternal origins of those with elongated skulls actually come from the Caucus regions.

> *"This brings us to the dumbfounding findings of the DNA tests: The bizarre DNA relationships that these skulls have to ancient and modern European populations. The only identifiable genetics in these skulls pertain to Western Europe, Northern Europe and the Middle East. None of the haplogroups of the indigenous populations of the Americas were detected in these DNA samples..."*

> *"....The only haplogroup detected was T2b, which occurs from the Middle East to Iceland, with heaviest concentrations in Georgia, Eastern Italy, Sardinia, Netherlands,*

Southern Portugal and Iceland (!) The broader T2 haplogroup is most heavily concentrated in the Eastern Baltic areas." [2]

Given this counterintuitive DNA evidence, I chose to use (mean) tissue depths based on studies of average-weight Armenian, Abkhazian, and Caucasian (female) ancestry instead of those with South American ancestry.

DESCRIPTION : 2,000+ YR OLD FEMALE ELONGATED SKULL - ORIGIN: PERU

Observations and indicators prior to reconstruction - (PS's added after reconstruction)

OVERALL SKULL SHAPE	Elongated skull. Long narrow Dolichocephalic-type face. Noted asymmetries, with most features on the left side of the subject's face, higher than the right side. In profile, from the glabella to the vertex is approx. 150mm. From the vertex to the most posterior point of the skull is approximately another 55mms. From the glabella to the vertex, the skull is smooth, sloping back at an approx 45deg angle. It is undulating toward the coronal suture with a rounded area of approx 65mm before the coronal suture. The skull then dips, creating a saddle-like area before rounding over the back of the skull toward the lambdoid suture. There is an anomalous foramen in the back of the skull which would normally fall along the line of the sagittal suture. However, there is no evidence of a sagittal suture. The occipital bone from the lambdoid suture down to the foramen magnum is elongated, quite flat and measures about 118mm from the suture to the middle of the foramen magnum. PS An approx 5mm was used to cover the skull itself, where no other depth markers were indicated.
FOREHEAD	The overall forehead is exceptionally long. Estimating the location of the hairline (widow's peak) the forehead is approx 49mm from the glabella **PS No hair has been indicated on this subject.**
BROW SHAPE	Not much brow ridge. Slight bony structure medial to the mid-supraorbital point, but no suggestion that the brow would be overhang the orbit.
EYE FISSURE	Slight 'diamond' shaped orbits, with a curved lower margin. Left orbit is slightly smaller overall than the right.

[2] https://myemail.constantcontact.com/DNA-Results-of-Elongated-Paracas-Skulls-from-Peru.html?soid=1108369064136&aid=WeUllPurSqQ

EYEBALL	Using a 25mm ball as a guide, it appears as though the eyes will be slightly protruding. Lining up the lacrimal crest and malar tubercle, the eyes will also be slightly slanted to be raised on the outer margins. **PS - The eye depth accommodated for a snug fit for a 23mm eyeball. I chose a dark hazel eye colour, as DNA testing has placed the origin of some of these skulls as coming from the Caucasus regions. It has also been noted that many of the elongated skulls were found with red hair. The eyes were centred in the orbit and the eyeballs were lined up with the iris slightly protruding from a line carried from the supraorbital ridge to the suborbital boney structure.**
EYELID PATTERN	With a thicker lateral rim, it is estimated that the eyelid fold will be more laterally defined. **PS An eyelid fold that was centrally uniform - from the lateral to medial - over the eyes seemed more natural.**
EYEBROW PATTERN	Eyebrows will be on the surface of the brow ridge (as opposed to tucked under the ridge), arched, and with a downward incline at the lateral ends, and not particularly long.
NASAL PROFILE	Not much projection at nasion. Relatively flat between glabella and end of nasal. The end of nasal has a slight rounding downward, suggesting a slight 'hook', but certainly not a strong 'hook'. There is a slight asymmetry, with the nose bending toward the subject's right side of skull, indicating the nose may not be perfectly straight. **PS If the eyes are positioned correctly, the nasal spine exhibited more height, and with a more prominent nose than I would have anticipated prior to reconstruction.**
NASAL WIDTH	The nasal aperture (piriform aperture) is 24mm, meaning that the overall with should be approx 33mm. There is indication of a deviated septim, which could affect the overall external look. The nasal opening is higher on the left side and slightly smaller.
NASAL SPINE	Long overall exterior projection. 7.5mm from volmer to tip of nasal spine. The nasal tip projection is straight with a total of 22.5mm on top of the philtrum depth. PS Given the straightness of the nasal spine (from the volmer), I could not see that the nose would be very hooked or rounded at the tip. The result is an attempt to match the indication of a slight 'hook' from the end of nasal, and marry it with the projection of 33mm (based on Krogman's formula of 3X length of nasal spine on top of Midphiltrum tissue depth).
ALAR SHAPE AND POSITION	Alar position will be slightly higher and slightly smaller on left side.
NOSTRIL POSITION	The nasal position will be evident in the Frankfort plane
TEETH	The teeth are relatively small, peg-like and well-worn. Eleven (11) of the 32 teeth are missing, leaving 21 teeth in the skull. Three wisdom teeth remain, which have erupted and show wear. At least 4 teeth seem to be lost prior to death, as the sockets have reformed bone. Others may have been lost postmortem or close to the death date. Given the over all wear, that the wisdom teeth are present (with wear), and that other missing teeth prior to death that have reformed bone, I am guessing that the subject could be between 40-60yrs old.

95

LIP THICKNESS	The upper teeth (maxillary) show some prognathism, which may indicate a slight 'overbite'. This means the upper lip will protrude over the bottom lip, even with the jaw 'slack', and both the upper and lower teeth may be slightly visible. The teeth are relatively small, indicating that both the upper and lower lips may be on the thinner side.
PHILTRUM WIDTH	Approx 8mm
MOUTH WIDTH	Approx 40mm - measured from incisors.
MOUTH CORNER INCLINATIONS	The markings for the levator anguli oris seem more pronounced that the depressor anguli oris, indicating that mouth corners may turn up instead of down. However, age may play a part in more downturned corners (marionette lines). PS reconstruction shows mouth turned down to indicate a more advanced age.
LIP SHAPE	The upper enamel line is relatively straight, indicating a rather flat lip line with only a slightly depressed philtrum.
NASOLABIAL FOLD	Fairly deep canine fossa - could indicate nasolabial fold. May also indicate a more advanced age. PS Following tissue depth markers, the nasolabial fold was minimal in the 'younger' version, but more pronounced in the 'aged' version.
CHIN SHAPE	The chin seems quite angular, and asymmetric at the mental eminence - the right side be significantly lower and more pointed than the left side. The mandible angle seems very square (for a female) with a rounded ramus.
EAR	The supra mastoid crest is smooth and the mastoid process is rough, indicating the ear may protrude more at the bottom than the top. The mastoid process is pointing downwards, indicating that the earlobe may be attached. I do not think the mastoid process is excessively large, so I will use more feminine, smaller ears. PS Note that I did not fashion the ears to protrude more at the bottom, nor did I attach the earlobes. This was not an intensional decision, but one of intuition. I did make the earlobes longer in the 'aged' version of the reconstruction.
CHEEK SHAPE	The zygomatic arches are thin and delicate, diminishing in size as they merge toward the temporal junction. Despite the 'masculine' angle of the mandible, and the delicate, thin zygomatic arch, I am wondering if the masseter might be less pronounced - adding to a look of 'high cheekbones'? The right side is slightly larger than the left, though the left side seems to have more lateral prominence than the right. The overall width of the face is 120mm at the zygion, somewhat larger than the overall width of the skull at 115mm. PS Though the zygomatic arch is fairly wide, it is not wider than the ears, making the ears appear quite flat against the head from a frontal view.

Tissue Depths:

Average Caucasoid / European Female (mean)

		Facial Points		Mean -Female	Armenian	Abkhazian	Average Caucasoid	Av Amer European
1	O	supraglabella		4.8	4.9	4.6	5.2	5.2
2	G	glabella		5.6	5.7	5.4	5.9	5.9
3	N	nasion		5.6	5.7	5.4	5.5	7
4	RHI	end of nasal		3.2	3.4	3	2.7	2.4
5	MP	mid-philtrum		9.9	10.1	9.7	8.5	12.3
6	LS	upper lip margin		10.4	10.8	10	9	11
7	LI	lower lip margin		12.1	12.2	11.9	10	12.6
8	LM	labiomental - lip fold		11.0	10.4	11.5	9.5	11.6
9	M	mental eminence - pogonion		11.1	10.8	11.3	10	11.3
10	MN	(menton - beneath chin) - gnathion		6.3	6.3	6.2	5.7	9.6
11	FE	lateral forehead	X2	5.4	5.7	5	3.5	4.9
12	OS	(mid)-supraorbital	X2	5.6	5.8	5.4	7	6.8
13	OR	orbitale - sub	X2	5.5	5.5	5.5	6	6.4
14	ZA	zygomaxillare - inferior malar	X2	11.3	12.3	10.3	12.7	10.8
15	LO	lateral orbit	X2	5.1	5.2	5	7.4	7.4
16	ZY	lateral zygomatic - mid	X2	9.0	8.9	9	7.5	9
17	SG	zygomatic - supraglenoid	X2	5.3	5.3	5.2	8	6.2
18	GO	gonion	X2	8.6	5.5	11.7	12	11.8
19	1M	supra M2 molar	X2	20.4	19.2	21.5	19.25	20.1
20	MM	midmasseter - occlusal line	X2	17.8	17.2	18.3	17	17
21	2M	sub M2 molar	X2	16.7	14.3	19	15.5	16.4
		Source - Helmer		Unknown Age	Lebedinskaya - 1993		Kollyman Buchly 1898	compilation (Wilkinson)

✳ All tissue depth measurements are in millimetres.

Reconstruction process

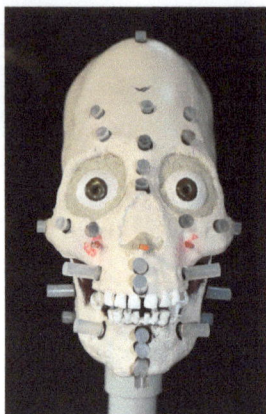

Depth markers and eyes in place

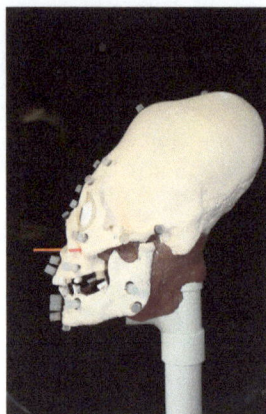

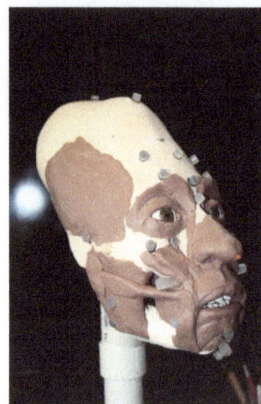

Application of musculature

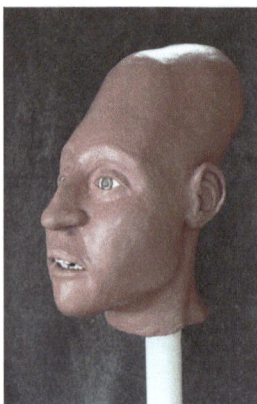

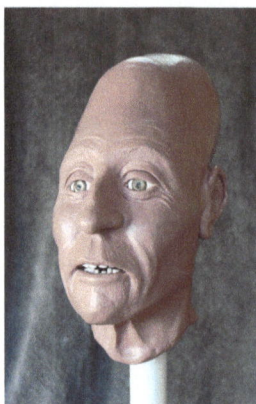

First reconstruction

Revision work

Unhappy with the first rendition, I decided to make changes. I replaced the subject's eyes with plastic replicas and aged the face further. To better portray gender, I added hair, headbands and earrings (common adornment for that period).

The other unique features, such as the asymmetry of the jawline and the crookedness in the nose, remained visible, just as indicated in the skull and identified in the observational reconstruction notes.

Final result

African/American female skull

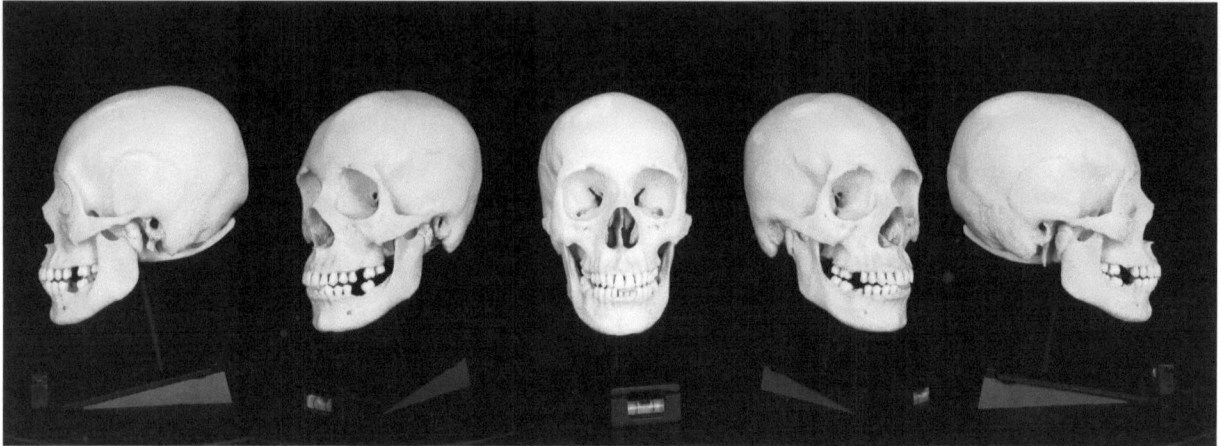

The subject of this reconstruction project was the skull of an African/American Female.

People of mixed European and African heritage, also known as Afro-Europeans or Afro-Eurasians, have a diverse anatomical makeup that reflects their unique genetic heritage.

Depending on the specific ancestry of the individual, they may exhibit physical traits commonly associated with both European and African populations, such as a range of skin tones, hair textures, and facial features. It is important to note that there is no "typical" physical appearance for Afro-Europeans, as each individual's genetic makeup is unique and can result in various physical features.

Therefore, the reconstruction artist must follow the lead of the information imparted by the characteristics of the skull itself. Because this skull was nowhere near the Frankfurt Plane on the stand that accompanied the skull itself, and I was having trouble securing the mandible, which was loose, I chose to connect the skull to a different armature made from PVC piping for the duration of the reconstruction process.

AFRICAN/AMERICAN FEMALE SKULL

Feature	Pre-construction observations and indicators - skull observed in the Frankfort Plan. Additional information from Bone Clone's report are added in Red. My additional comments may be added to notes (in bold) after reconstruction.
General Observations	African/American Female. This individual appears to be >50, based on (my limited knowledge of) dentition (loss of teeth and remodelled bone - premortem and postmortem) and (though a controversial method) an almost total fusion of the coronal, lambdoid, and sagittal sutures. This individual displays both African and American features, with the more rectangular orbital orifice, but with a more narrow nasal opening. This individual seems fairly symmetrical, with the exception of a slightly heavier right-side mandible, and eye-socket. Whilst the right-side zygomatic bone is heavier, it is more receding when viewed from above. It is noted that the left styloid process has been broken off. This could have been postmortem or in the casting process. The left styloid process is much shorter than the right and has an irregular, exostotic thickening at its termination (healing/healed fracture). Cranial vault (landmarks 1-7) 48.8 =/-10.5 years. Anterior cranium vault (landmarks 6-10) 43.4 +/-10.7 years. 38.3 - 54.1 years.
OVERALL SKULL SHAPE	The distance from the glabella to the external occipital protuberance is approx 168mm. The distance widest breadth is approx 134mm, with a cephalic index of 79.8* This index makes for a **mesaticephalic** skull that is nearly oval. In profile, from the glabella to the vertex is approx. 100mm. The skull dips slightly after the coronal suture (post-bregmatic depression). The cranial vault is nicely rounded from the lateral view. From the posterior view, the cranial vault is somewhat 'triangular' - sloping from the sagittal suture down the parietal bones to the superior temporal linea, then nearly straight down to the mastoid processes. There are two anomalous foramen in the back of the skull which fall along the line of the sagittal suture and above the area of the posterior fontanelle. The external occipital protuberance measures about 52mm to the posterior edge of the foramen magnum. The calvarial sutures are complex.

FOREHEAD	From the glabella to the vertex is approx 100mm. From the vertex to the most posterior point of the skull is approximately another 70mms. From the glabella to the vertex, the skull has a slight protruding 'ridge' that runs vertically along the metopic suture. Noting the change from 'smooth' skull to a more grainy texture, I estimate that the hairline starts approx 70mm from the nasion (measured with callipers).
BROW SHAPE	Flat vertical brow ridge (sideview). Slight bony structure medial to the mid-supraorbital point, but no suggestion that the brow would be overhang the orbit. The attachments for the Corrugator Spericilii and Procerus muscle groups do not appear to be prominent, suggesting that she may have had a smooth brow. Note that the R.Supraorbital Notch (foramen) is either broken peri or postmortem, or in the casting process, or hasn't formed into a typical foramen 'hole', as it has on the left side. Mild prominence of supraorbital tori attachment. The supraorbital margins are sharp.
EYE FISSURE	Slight 'diamond' shaped orbits, with a curved lower margin. Left orbit is slightly smaller overall than the right. The horizontal measurement for the right side opening is 3mm larger than the left, and 2mm greater in the vertical. (V-R40mm V-L38mm, H-R41mm H-L41mm). Interocular distant is slightly widened.
EYEBALL	Using a 25mm ball as a guide, it appears as though the eyes will be 'average' depth within the orbit cavity, with the right eye being slightly forward of the left. However, given a deep eyesocket, and estimating that the age is >50yrs, the eyes may have a more sunken appearance. Lining up the lacrimal crest and malar tubercle, the right eye will be slightly more raised than the left side.
EYELID PATTERN	With a thicker lateral rim, it is estimated that the eyelid fold will be more laterally defined. R.eye seems to have slightly more medial supraorbital overhang. The right eye may 'droop' more in the centre, whilst the fold downward 'slant' may be more pronounced on the left side
EYEBROW PATTERN	Eyebrows will be on the surface of the brow ridge (as opposed to tucked under the ridge), arched, and with a downward incline at the lateral ends. See photo example.

NASAL PROFILE	Nasal projection is more pronounced than her African male counterpart, which may be an indication of her European ancestry. The nasion is straight. The nasal spine itself measures 7.5mm from volmer to tip. A total of 22.5mm projection will be added to the philtrum depth of approx 11.5mm (34mm).** Lebedinshaya method - 28mm projection.
NASAL WIDTH	The overall shape of the opening is an elongated, upside-down heart shape. The overall width, taken at the nasion, just above the lacrimal crest, is 22mm. The nasal aperture (piriform aperture) is 27mm at its widest, indicating overall width between 37-43mm - taking into consideration mixed ancestry. There is indication of a very slight deviated septum (to the right side), but I don't envisage an affect on overall external nose position. The base of the nasal opening is higher on the left side and slightly smaller.
NASAL SPINE	The nasal spine is sharp, defined, and slightly bending to the right. The spine itself is slightly downturned, whist the underside points 'straight' forward. The underside of the nasal spine carries on downward with a prominent vertical thin column of bone running down the maxilla/philtrum for approx 9mms before disappearing into the bone above the central incisors (see photo below). The nasal spine itself measures 7.5mm from volmer to tip.
ALAR SHAPE AND POSITION	Alar position will be slightly higher and slightly smaller on left side. No notable alar 'shelf' or nasal guttering. Very slight suggestion of a right gutter.
NOSTRIL POSITION	The nasal position is expected to be seen (in the Frankfort plane) once tissue has been added
TEETH	Overall, the teeth seem be in good condition, with little wear, and appear to be 'average' in size.*** Five (5) of 32 teeth are missing, leaving 27 teeth in the skull. Three erupted wisdom teeth are remaining (lower R missing). Two (2) teeth seem to be lost antemortem (1M left side upper and lower), as the sockets have reformed bone. Two (lower L wisdom & upper R1M) may have been lost postmortem or perimortem, as the bone has not reformed. One tooth (1M, lower/right) has been broken off at the gum-line. It would be difficult for me to determine when. The lack of wear either denotes a younger age at death (than I estimate), or perhaps a softer 'western' diet. Relatively straight occlusion line. Mild degree of attrition.

LIP THICKNESS	The upper (maxillary) and lower (mandibular) teeth show slight prognathism. The teeth , indicating that both the upper and lower lips may be on the fuller side.
PHILTRUM WIDTH	Approx 10mm - Incisor angle suggest that the philtrum may be more triangular, as opposed to rectangular.
MOUTH WIDTH	Approx 40mm - measured from incisors.
MOUTH CORNER INCLINATIONS	The markings for the levator anguli oris seem more pronounced than the depressor anguli oris, indicating that mouth corners may turn up instead of down. However, age may play a part in more downturned corners (marionette lines).
LIP SHAPE	The upper enamel line is relatively horizontal. Given the African ancestry, the lips may be full. With the bony spine running vertically down the philtrum, meeting the vermillion line, might make the upper lip seem more turned up.
NASOLABIAL FOLD	Fairly deep canine fossa - could indicate nasolabial fold. May also indicate a more advanced age.
CHIN SHAPE	The chin is rounded with a strong mental eminence. The right side of the chin is slightly more robust than the left side. The mandible angle is 120degs, with a rounded ramus and slight flair at the gonion.
EAR	The supra mastoid crest is smooth and slightly protruding. The mastoid process is rough and can be seen when viewing face on, indicating the ear may be easily visible and somewhat proturding. The mastoid process is pointing forwards, indicating that the earlobe may be unattached.**** The ear will be angled to reflect the 120degs of the mandible.
CHEEK SHAPE	The zygomatic arches are not significantly robust, with a breadth of 122mm, slightly less than that of the cranium. The zygomatic arch itself is straight and symmetrical in shape (elongated triangular shape) as it merges toward the temporal junction. The right side of the zygomatic bone is more recessed from the frontal plane, but slightly more protruding at the lower corner of the orbital bone - laterally - than the left side.
References:	

* Cephalic Index used to determine dolichocephalic (long headed), mesaticephalic (moderate headed), or brachycephalic (short headed) cephalic index or cranial index. Anders Retzius (1796–1860) An index < 75 - **dolichocephalic**. An index of 75-80 - **mesaticephalic**. An index >80 - **brachycephalic**.

** Width - (European): aperture of 3/5ths the overall nose (Gerasimov). (Negroid): aperture measured at its widest point; 16mm added for total width (8mm added to each side of aperture), (Krogman - Karen Taylor). **Alternately**: The profile of the nose as determined by the Lebedinskaya method. Line A dissects the nasion and prothion. Line B is parallel to line A and intersects the foremost point on the nasal bone. Four to six equidistant lines of the piriform aperture. At each perpendicular line, the distance from line B to the piriform rim was measured and the same distance added on the other side of the line B. Using this method the profile of the nose was predicted. modified from Prokopec and Ubelaker (2002).

*** Dr Welfel (1974-1974) study of tooth size - averages

**** Fedosyutkin and Nainys (1993) estimates of ear positioning.

African/American Female Tissue Depths

Average African/European Female (mean)

		Facial Points		Average (mean)	(F) African/ American	South African Female	Female Amer European
1	O	supraglabella		4.79	4.5	4.68	5.2
2	G	glabella		6.03	6	6.2	5.9
3	N	nasion		6.08	5.25	6	7
4	RHI	end of nasal		2.96	3.75	2.72	2.4
5	MP	mid-philtrum		11.49	11.25	10.92	12.3
6	LS	upper lip margin		12.27	12.5	13.3	11
7	LI	lower lip margin		14.08	15	14.65	12.6
8	LM	labiomental - lip fold		12.02	12.25	12.21	11.6
9	M	mental eminence - pogonion		11.47	12.5	10.6	11.3
10	MN	(menton - beneath chin) - gnathion		8.11	8	6.72	9.6
11	FE	lateral forehead	X2	4.55	4	4.75	4.9
12	OS	(mid)-supraorbital	X2	7.21	8	6.84	6.8
13	OR	orbitale - sub	X2	7.18	8.25	6.89	6.4
14	ZA	zygomaxillare - inferior malar	X2	15.41	16.75	18.67	10.8
15	LO	lateral orbit	X2	11.85	13		10.7
16	ZY	lateral zygomatic - mid	X2	8.97	9.5	8.41	9

		Facial Points		Average (mean)	(F) African/ American	South African Female	Female Amer European
17	SG	zygomatic - supraglenoid	X 2	9.90	11.5	12.01	6.2
18	GO	gonion	X 2	14.40	13.5	17.9	11.8
19	1M	supra M2 molar	X 2	23.49	20.25	30.11	20.1
20	MM	midmasseter - occlusal line	X 2	19.28	19.25	21.6	17
21	2M	sub M2 molar	X 2	18.36	17	21.67	16.4
		Source		Age unknown	Rhine & Campbell	Cavanagh & Steyn 2011	compilation (Wilkinson)

✳ Being of mixed ancestry, I used an average of several different ancestry tissue depth studies.

✳ All tissue depth measurements are in millimetres.

Reconstruction process

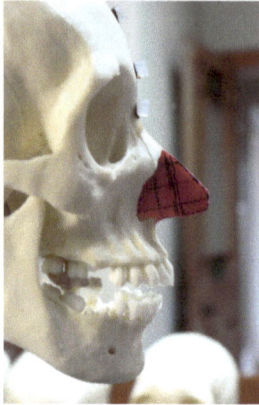

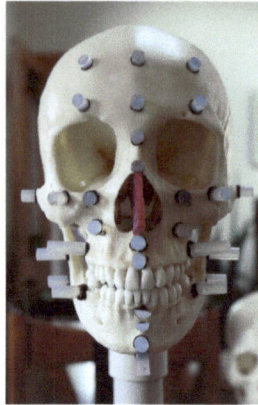

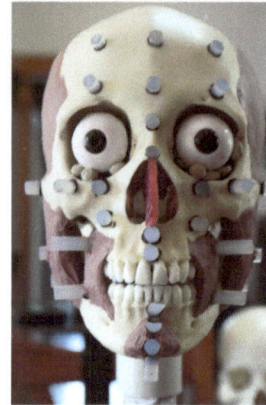

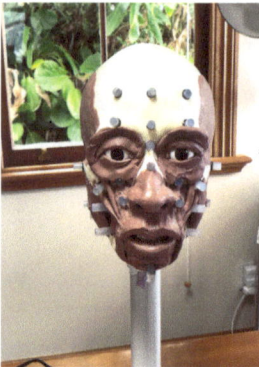

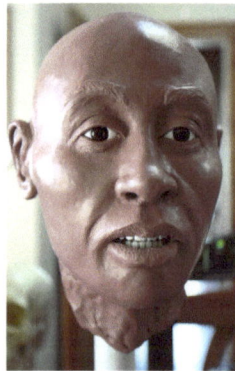

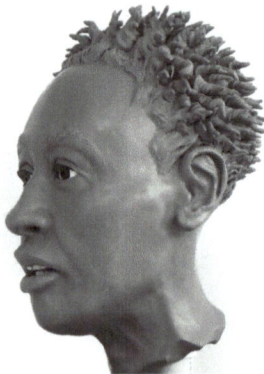

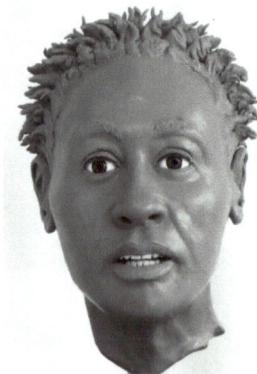

African Male Skull

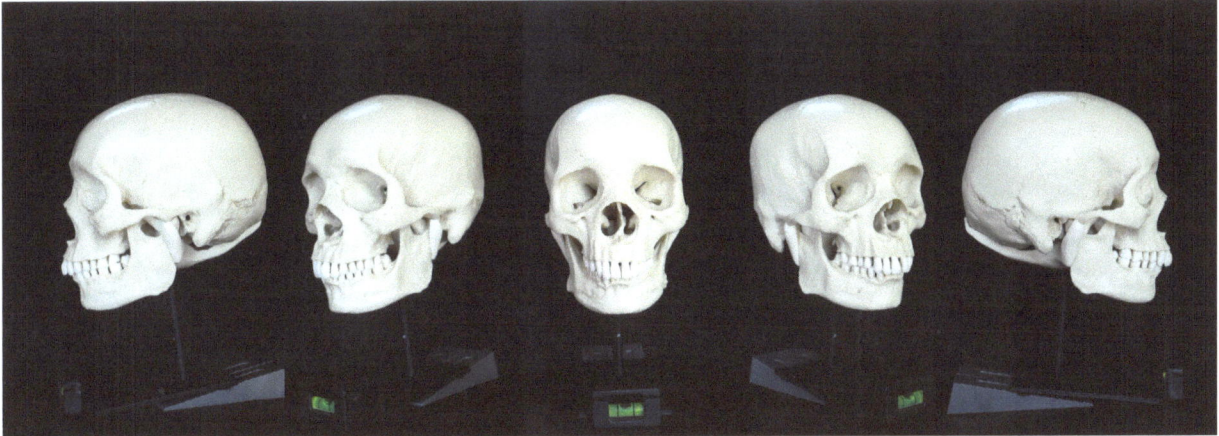

The morphology of the African skull varies significantly due to the continent's diverse population and complex history of human evolution. However, some general features are commonly associated with the African skull. For example, the African skull tends to be more robust than other human populations, with thicker cranial bones and a more powerful facial structure. These features have been linked to adaptations for chewing tough foods and resisting the forces of the jaw muscles during mastication.

Another characteristic of the African skull is a more pronounced brow ridge, which is thought to have evolved as a protection for the brain during hunting and other activities that involve impact to the face. In addition, many African populations have a more prognathic or forward-projecting jaw, which is believed to be an adaptation for a diet that includes tough and fibrous foods. These features may also contribute to the development of a larger nasal cavity, which is thought to be an adaptation to hot and arid environments.

Overall, the African skull's morphology reflects the continent's diverse and complex evolutionary history. While some general features are associated with African populations, many variations exist within and among different people. As our understanding of human evolution continues to evolve, we will likely gain new insights into the factors that have shaped the morphology of the African skull and other aspects of human biology.

AFRICAN MALE SKULL

Features	Pre-construction observations and indicators - skull observed in the Frankfort Plan. Additional information from Bone Clone's report are added in Red. My additional comments may be added to notes (in bold) after reconstruction.
General Observations	African Male. This individual appears to be <50yrs, perhaps between 35-45yrs, based on (my limited knowledge of) dentition. This individual displays typical African features, with the more rectangular orbital orifices, with a broad nasal opening. This individual seems fairly symmetrical, with the exception that the right-side of the maxilla is slightly more protruding (when viewed from above). It is noted that both styloid processes have broken off. This could have been postmortem or in the casting process.
OVERALL SKULL SHAPE	The distance from the glabella to the external occipital protuberance is approx 179mm. The distance widest breadth is approx 129mm, with a cephalic index of 72.* This index makes for a **dolichocephalic** long skull (elongated in the anteroposterior plane). In profile, from the glabella to the vertex is approx. 116mm. The cranial vault is nicely rounded from the lateral view. From the posterior view, the cranial vault is squarish - sloping from the sagittal suture down the parietal bones, with the most lateral protrusion in the centre of the parietal bones, then slightly curving in, down to the mastoid processes. The sagittal suture is quite evident at the juncture of the posterior fontanelle. The external occipital protuberance measures about 46mm to the posterior edge of the foramen magnum.
FOREHEAD	From the glabella to the vertex is approx 110mm. From the vertex to the most posterior point of the skull is approximately another 75mms. From the glabella to the vertex, the skull is uniformly rounded. Noting the change from 'smooth' skull to a more grainy texture, I estimate that the hairline starts approx 72mm from the nasion (measured with callipers). The glabella is 'depressed'.

BROW SHAPE	Flat vertical brow ridge (sideview). Bony structure medial to the mid-supraorbital point, with a suggestion that the brow would be overhang the orbit (centrally). The attachments for the Corrugator Spericilii and Procerus muscle groups appear to be prominent, suggesting that there may be brow 'wrinkles'. Note that both Supraorbital Notch (foramen) have not formed into a typical foramen 'hole'.
EYE FISSURE	Slight 'diamond' shaped orbits, with a curved lower margin. Left orbit is slightly smaller overall than the right. The horizontal measurement for the right side opening is 1.5mm larger than the left, but the left is 1mm greater in the vertical than the right. (V-R34mm V-L34mm, H-L40mm H-R38mm).
EYEBALL	Using a 25mm ball as a guide, it appears as though the eyes will be 'average' depth within the orbit cavity, with the right eye being slightly posterior of the left. Lining up the lacrimal crest and malar tubercle, the right eye will be slightly lower than the left side by approx 2mm. The interocular distance is broad.
EYELID PATTERN	With a thicker lateral rim, it is estimated that the eyelid fold will be more laterally defined. R.eye seems to have slightly more medial supraorbital overhang. The right eye may 'droop' more in the centre, whilst the fold downward 'slant' may be more pronounced on the left side
EYEBROW PATTERN	Eyebrows will be tucked under the ridge and slightly curved. See example photo.
NASAL PROFILE	Nasal projection is flattened and short (15mm from nasion to end of nasion). The nasion is straight. The nasal spine itself measures 11mm from volmer to tip. The projection will be assessed at time of reconstruction, as the standard 'formula' would make the projection 45mm (33mm + philtrum depth of approx 12mm) (3Xnasal + average philtrum depth).** This seems excessive and alternate methods may be applied.

NASAL WIDTH	The overall shape of the opening is large and rounded. The overall width, taken at the nasion, just above the lacrimal crest, is 26mm. The nasal aperture (piriform aperture) is 34mm at its widest, indicating overall width of approx 50mm. There is evidence of a deviated septum (to the left side), but as there is no dosplacement of the nasal spine, I don't envisage an affect on overall external nose position. The base of the nasal opening is higher on the left side, and smaller by approx 2mm laterally.
NASAL SPINE	The nasal spine is well defined. The spine points 'straight' forward. The nasal spine itself measures 11mm from volmer to tip. The nasal root is depressed. The anterior nasal spine is short.
ALAR SHAPE AND POSITION	Alar position will be slightly higher and slightly smaller on left side. Very notable alar 'shelf' or nasal guttering (see photo below).
NOSTRIL POSITION	The nasal position (in the Frankfort plane) may be hidden once tissue has been added
TEETH	Overall, the teeth show wear, and appear to be larger than 'average' in size.*** 30 of 32 teeth are present. It appears that neither of the upper 3Ms (on both sided) have never erupted, otherwise, the teeth are in good condition. Supraeruption of 3.8 [#17] and 4.8 [#32]. The wear seems to indicate an age of 30-45yrs. Cranial vault (landmarks 1-7) 39.4 +/-9.1yrs. Anterior cranium (landmarks 6-10) 45.5 +/- 8.9 years). Estimate age 36.6-48.5 years. Relatively straight occlusion line, barring the elongated lower 3Ms, due to no 3Ms above. The maxilla is (alveolar) prognathic, with the left central incisor more projected that the right. The dental arcade is somewhat rectangular.
LIP THICKNESS	The upper (maxillary) teeth show prognathism - the upper lip may project further forward than the left. The larger teeth indicate that both the upper and lower lips may be on the fuller side.
PHILTRUM WIDTH	Approx 11mm - Incisor angle suggest that the philtrum may be more square. Subnasal prognathism
MOUTH WIDTH	Approx 40mm - measured from incisors.

MOUTH CORNER INCLINATIONS	The markings for the levator anguli oris seem more pronounced than the depressor anguli oris, indicating that mouth corners may turn up instead of down.
LIP SHAPE	The upper enamel line is relatively horizontal. Given the African ancestry, the lips may be full. Given the subnasal prognathism, the upper lip may be more pronounce.
NASOLABIAL FOLD	Fairly deep canine fossa - could indicate nasolabial fold. May indicate more rounded and prominent cheeks.
CHIN SHAPE	The chin is square and robust,. The chin is slightly receded. The mandible angle is 90degs, with a rounded ramus and strong flair at the gonion.
EAR	The supra mastoid crest is smooth and sharply protruding. The mastoid process is rough and can notbe seen when viewing face on, indicating the ear may be easily visible and somewhat flat to the side of the head. The mastoid process is pointing forwards, indicating that the earlobe may be unattached.**** The ear will be angled to reflect the 90degs of the mandible.
CHEEK SHAPE	The zygomatic arches are not significantly robust, with a breadth of approx 124mm, equal to the breadth of the crainim. The zygomatic arch itself curves up and then down as it merges toward the temporal junction. The right side of the zygomatic bone is more recessed from the frontal plane, and the left side is slightly more protruding at the lower corner of the orbital bone.
References:	
	* Cephalic Index used to determine dolichocephalic (long headed), mesaticephalic (moderate headed), or brachycephalic (short headed) cephalic index or cranial index. Anders Retzius (1796-1860) An index < 75 - **dolichocephalic**. An index of 75-80 - **mesaticephalic**. An index >80 - **brachycephalic**.

** Width - (European): aperture of 3/5ths the overall nose (Gerasimov). (Negroid): aperture measured at its widest point; 16mm added for total width (8mm added to each side of aperture), (Krogman - Karen Taylor). **Alternately**: The profile of the nose as determined by the Lebedinskaya method. Line A dissects the nasion and prothion. Line B is parallel to line A and intersects the foremost point on the nasal bone. Four to six equidistant lines of the piriform aperture. At each perpendicular line, the distance from line B to the piriform rim was measured and the same distance added on the other side of the line B. Using this method the profile of the nose was predicted. modified from Prokopec and Ubelaker (2002).

*** Dr Welfel (1974-1974) study of tooth size - averages

**** Fedosyutkin and Nainys (1993) estimates of ear positioning.

Tissue depth studies for African males

Finding tissue depth studies for African males proved more difficult than expected. I only found a simplified 'Zulu' study for African men, giving scant information.

Instead, I had to improvise to come up with some averages. I used American studies and adjusted average differences calculated between males and females (using South African female data).

Average African Male (mean)

		Facial Points		Average (mean)	Zulu	American Negroid	Adjusted male/ female deviation	Black Male American?	Adjusted male/ female deviation	Female -South African	Gujarati Male/ Female deviation
1	O	supraglabella		4.97	5.21	5	4.93	4.75	4.93	4.68	0.25
2	G	glabella		6.28	5.21	6.25	7.41	6.25	7.41	6.2	1.21
3	N	nasion		5.74	5.21	6	5.74	6	5.74	6	-0.26
4	RHI	end of nasal		3.40	3.8	3.75	2.27	3.77	2.27	2.72	-0.45
5	MP	mid-philtrum		11.98	12	12.25	11.4	12.25	11.4	10.92	0.48
6	LS	upper lip margin		13.88		14.25	13.4	14	13.4	13.3	0.1
7	LI	lower lip margin		15.30		15.5	15.4	15	15.4	14.65	0.75
8	LM	labiomental - lip fold		12.07		11.75	12.45	12	12.45	12.21	0.24
9	M	mental eminence - pogonion		11.20		11.5	9.85	12.25	9.85	10.6	-0.75
10	MN	(menton - beneath chin) - gnathion		7.49		8.25	6.22	8	6.22	6.72	-0.5
11	FE	lateral forehead	X 2	4.88		5	4.75	4.9	4.75	4.75	
12	OS	(mid)-supraorbital	X 2	7.78		8.50	7.35	7.5	7.35	6.84	0.51

	Facial Points		Average (mean)	Zulu	American Negroid	Adjusted male/female deviation	Black Male American?	Adjusted male/female deviation	Female-South African	Gujarati Male/Female deviation
13	OR	orbitale - sub	X2	7.65		7.75	7.44	7.75	7.44	6.89	0.55
14	ZA	zygomaxillare - inferior malar	X2	17.28		16.5	18.33	17	18.33	18.67	-0.34
15	LO	lateral orbit	X2	13.25		13.25		13.25			
16	ZY	lateral zygomatic - mid	X2	8.45		8.25		8.65		8.41	
17	SG	zygomatic - supraglenoid	X2	11.38		11		11.75		12.01	
18	GO	gonion	X2	15.22		13	18.4	14.25	18.4	17.9	0.5
19	1M	supra M2 molar	X2	25.16		23	30.38	22.1	30.38	30.11	0.27
20	MM	midmasseter - occlusal line	X2	19.00		19		19		21.6	
21	2M	sub M2 molar	X2	18.44		16.5	23.07	15.75	23.07	21.67	1.4
		Source		Age Unknown	Aulsebrook WA1, Becker PJ, Işcan MY (1996)	Rhine & Campbell	Anand Lodha, Mitalee Mehta, M.N. Patel, Shobhana K. Menon (2016)	unattributed internet source - http://what-when-how.com/forensic-sciences/facial-tissue-thickness-in-facial-reconstruction/	Anand Lodha, Mitalee Mehta, M.N. Patel, Shobhana K. Menon (2016)	Cavanagh & Steyn 2011	Anand Lodha, Mitalee Mehta, M.N. Patel, Shobhana K. Menon (2016)

* All tissue depth measurements are in millimetres.

Reconstruction process

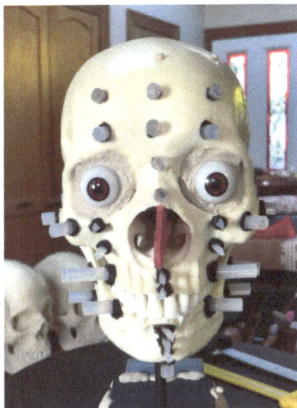

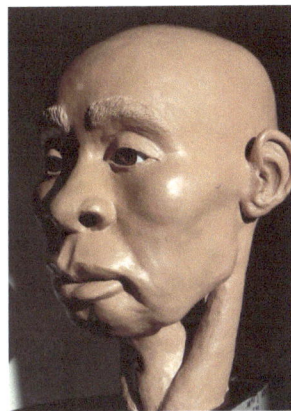

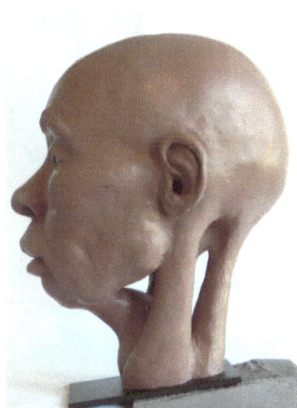

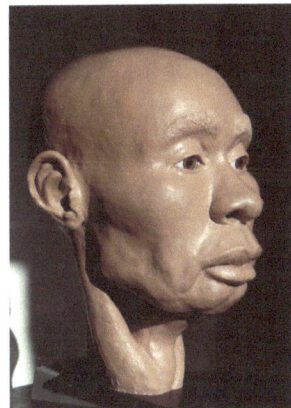

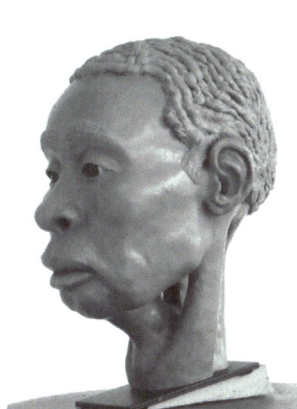

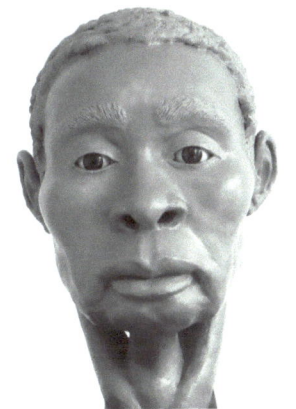

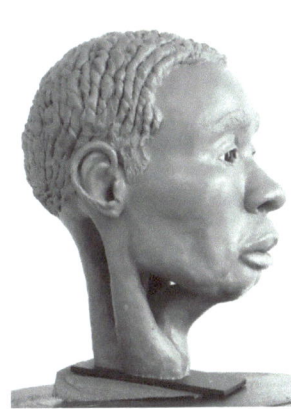

Asian female skull

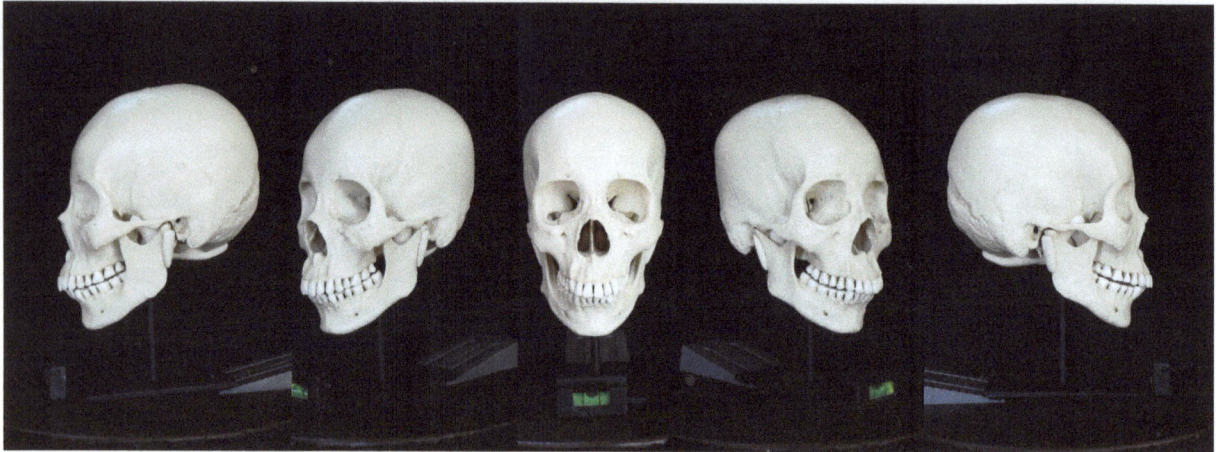

The morphology of the Asian female skull is distinct from that of other ethnic groups, with specific physical characteristics that set it apart. One of the most notable differences is the shape of the skull, which tends to be more rounded and shorter in length than in other populations. The forehead is also typically flatter, with less pronounced brow ridges and a lower forehead angle than in other groups.

In addition to these differences in skull shape, Asian females tend to have more delicate facial features, such as the nose, eyes, and mouth. The eyes are often described as "almond-shaped," with a slight upward tilt at the outer corners. The cheekbones are also less pronounced, with a flatter mid-face region.

Another distinctive feature of the Asian female skull is its smaller cranial capacity, meaning that the brain size of Asian females is typically smaller compared to other ethnic groups. However, it is vital to note that differences in cranial capacity do not necessarily indicate differences in intelligence or cognitive abilities. Einstein had a smaller-than-average brain!

Overall, the morphology of the Asian female skull is unique and reflects the genetic and evolutionary history of this population. While some physical characteristics are more common among Asian females, there is significant variation within and between all ancestral groups, and individual differences should be recognized and appreciated.

ASIAN FEMALE SKULL

Features	Pre-construction observations and indicators - skull observed in the Frankfort Plan. Additional information from Bone Clone's report are added in Red. My additional comments may be added to notes (in bold) after reconstruction.
General Observations	Asian Female. This individual appears to be <50, based on (my limited knowledge of) dentition and (though a controversial method) with some fusion of the coronal, lambdoid, and sagittal sutures. This individual displays typical asian features, with the more rounded orbital orifice, and a wide, flat nasion. This individual seems fairly symmetrical, with the exception of a slightly more 'flaired' right-side mandible. Both styloid processes are short. Cranial vault (landmarks 1-7) 45.2yrs +/-12.6 years. Anterior cranium (landmarks 6-10)56.2 =/- 8.5 years. Estimate 47.7 - 57.8 years.
OVERALL SKULL SHAPE	The distance from the glabella to the external occipital protuberance is approx 164mm. The distance widest breadth is approx 135mm, with a cephalic index of 79.3.* This index makes for a **mesaticephalic** skull. In profile, from the glabella to the vertex is approx. 113mm. The skull bulges slightly before the coronal suture. There is a slight post-bregmatic depression. The cranial vault is nicely rounded from the lateral view. From the posterior view, the cranial vault is somewhat 'boxy' - sloping from the sagittal suture down the parietal bones where there is a significant bulge, then curving back into the superior temporal linea, then gently curving out to the mastoid processes. The lambdoid suture on either side of the posterior fontanelle is depressed, wide and very visible.. The external occipital protuberance, which is not prominent, measures approx 47mm to the posterior edge of the foramen magnum.
FOREHEAD	The forehead is very flat and vertical with slightly protruding bone structure approx 25-35mm from the mid-supraorbital points. From the glabella to the vertex is approx 127mm. From the vertex to the most posterior point of the skull is approximately another 120mms. From the glabella to the vertex, the skull has a very slight protruding 'ridge' that runs vertically along the metopic suture. Noting the change from 'smooth' skull to a more grainy texture, I estimate that the hairline starts approx 56mm from the nasion (measured with callipers).

BROW SHAPE	Flat vertical brow ridge (sideview). Slight bony structure medial to the mid-supraorbital point, but no suggestion that the brow would be overhang the orbit. There is some suggestion of attachments for the Corrugator Spericilii and Procerus muscle groups, but not enough to suggest that she would have anything but a smooth brow. Note the R.Supraorbital Notch (foramen) is higher than on the left side.
EYE FISSURE	Slight 'rounded' shaped orbits, with a curved lower margin. Left orbit is slightly smaller overall than the right. The horizontal measurement for the right side opening is 3mm larger than the left, and 2mm greater in the vertical. (V-R33mm V-L33mm, H-R34 H-L36mm).
EYEBALL	The eyeball sockets are quite shallow compared to the African and European subjects. Using a 25mm ball as a guide, it appears as though the eyes will be 'protruding' somewhat within the orbit cavity. I may also mean I use a smaller eyeball size during reconstruction. The right eye will be slightly forward of the left. Lining up the lacrimal crest and malar tubercle, the right eye will be slightly more raised than the left side. The interocular distance is not significantly widened.
EYELID PATTERN	With a thicker lateral rim on the left side, the eyelid may 'dip' more on that side. Given the ancestry of this skull, I expect to produce a recognisable epicanthal (Mongolian) fold that will semi-obscure the lacrimal caruncle and be determined by the projection of the nasion.
EYEBROW PATTERN	Eyebrows will be on the surface of the brow ridge (as opposed to tucked under the ridge). I expect the brows to be superficial and slightly curved.
NASAL PROFILE	Nasal projection is very flat and shallow, with a tiny bumping then a round at the end of the nasion. The nasal spine itself measures approx 7mm from volmer to tip (though the tip is not well defined). A total of 27.5mm projection will be added to the philtrum depth of approx 10.1mm (37.6mm).** However, given ancestry, I will use various calculations (methods) and adjust the dimensions to fit the face, should the nose projection look out of place.

NASAL WIDTH	The overall shape of the opening is heart shape. The overall width, taken at the nasion, just above the lacrimal crest, is 21mm. The nasal aperture (piriform aperture) is 25.4mm at its widest, indicating overall width between 42-46mm - taking into consideration ancestry. The septum is straight, and there is some nasal guttering. The base of the nasal opening is slightly higher on the left side and
NASAL SPINE	The nasal spine is not well defined, but straight. The spine itself is slightly straight and any indication of projection is missing. The underside of the nasal spine carries on downward with a prominent vertical thin column of bone running down the maxilla/philtrum for approx 5.8mms before disappearing into the bone above the central incisors. The nasal spine itself measures approx 7mm from volmer to tip (though the tip is not well defined).
ALAR SHAPE AND POSITION	Alar position will be slightly higher and slightly more rounded on left side. Notable alar 'shelf' or nasal guttering.
NOSTRIL POSITION	The nasal position may not be evident (in the Frankfort plane) once tissue has been added
TEETH	Overall, the teeth seem be in good condition, with little wear, and appear to be 'average' in size.*** All 32 teeth are in the skull. All wisdom teeth have erupted. Wear is evident in both 1Ms, with some on both 2Ms, and little on the 3Ms. I would estimate age of 35-45yrs. Moderate degree of attrition on the occlusal surfaces of the dentition. The teeth show 'shovelling' - the posterior/interior of the incisors are 'scooped out'. Age range 32.6 - 64.7 years.
LIP THICKNESS	The upper (maxillary) and lower (mandibular) teeth show significant prognathism. The canines are particularly evident in the maxilla (maxilla vault), but do not protrude beyond the other teeth. I would expect flat, but 'full' lips, with evidence of an 'overbite'.
PHILTRUM WIDTH	Approx 10.8mm - Incisor angle suggest that the philtrum may be more triangular, as opposed to rectangular, and long.
MOUTH WIDTH	Approx 36mm - measured from incisors.

MOUTH CORNER INCLINATIONS	The markings for the levator anguli oris and the depressor anguli oris do not seem particularly prominent. Though, given the prognathism of the maxilla, I would expect the corners to be slightly upturned.
LIP SHAPE	The upper enamel line curves upward. Given the significant upper (maxillary) and lower (mandibular) prognathism, I would expect the lips to be protruding (perhaps 'pouting').
NASOLABIAL FOLD	Shallow canine fossa - making for a 'flatter', more rounded face.
CHIN SHAPE	The chin has little mental eminence. The right side of the chin is slightly skewed to the right and the gonion is more robust than the left side. The mandible angle is approx 120degs, with a rounded ramus and significant flair at the gonion. The mandibular deviation to the right will make the face look somewhat longer on the left side, and the jaw will be slightly right of the midline. The mandible is somewhat squared. Broad ascending mandibular ramus.
EAR	The supra mastoid crest is smooth and flat. The mastoid process is cannot be seen when viewing face on, indicating the ear may not be easily visible. The mastoid process is pointing forward slightly, indicating that the earlobe may or may not be unattached.**** The ear will be angled to reflect the 120degs of the mandible.
CHEEK SHAPE	The zygomatic arches are robust, with a breadth of 116mm. The zygomatic arch itself is fairly straight and with a slight curve upwards midway, as it merges toward the temporal junction. The right side of the zygomatic bone is more recessed from the frontal plane, and slightly less protruding at the lower corner of the orbital bone - laterally - than the left side, but slightly sharper.
References:	
	* Cephalic Index used to determine dolichocephalic (long headed), mesaticephalic (moderate headed), or brachycephalic (short headed) cephalic index or cranial index. Anders Retzius (1796-1860) An index < 75 - **dolichocephalic**. An index of 75-80 - **mesaticephalic**. An index >80 - **brachycephalic**.

** Width - (European): aperture of 3/5ths the overall nose (Gerasimov). (Negroid): aperture measured at its widest point; 16mm added for total width (8mm added to each side of aperture), (Krogman - Karen Taylor). **Alternately**: The profile of the nose as determined by the Lebedinskaya method. Line A dissects the nasion and prothion. Line B is parallel to line A and intersects the foremost point on the nasal bone. Four to six equidistant lines of the piriform aperture. At each perpendicular line, the distance from line B to the piriform rim was measured and the same distance added on the other side of the line B. Using this method the profile of the nose was predicted. modified from Prokopec and Ubelaker (2002).

*** Dr Welfel (1974-1974) study of tooth size - averages

**** Fedosyutkin and Nainys (1993) estimates of ear positioning.

Tissue depths

Asian Female (mean)

#		Facial Points		Average (mean)	Japanese Female	Chinese (NZ) F-25-30yrs	Chinese Female	Korean Female	Korean Female	Buryats Female	Asian-derived - SW Am. Indian - N Female
1	O	supraglabella		3.9	2.0		3.9	5.2			4.5
2	G	glabella		5.0	3.2	5.8	5.32	5.4	5.4	5.6	4.5
3	N	nasion		4.7	3.4			4.4	4.4	4.5	7
4	RHI	end of nasal		2.4	1.6		2.4	2.9			2.5
5	MP	mid-philtrum		10.1			10.64	9.6			10
6	LS	upper lip margin		10.9		10.4		10.6	10.6	11.7	11
7	LI	lower lip margin		12.5				12.3	12.3	13.1	12.25
8	LM	labiomental - lip fold		9.9	8.5			11.1			10
9	M	mental eminence - pogonion		10.3	5.3		9.12	11.1	11.1	11.9	13
10	MN	(menton - beneath chin) - gnathion		6.1	2.8	8	5.36	6.5			8
11	FE	lateral forehead	X2	4.3				4.5			4
12	OS	(mid)-supraorbital	X2	8.5							8.5
13	OR	orbitale - sub	X2	5.3	3.6		5.96				6.25
14	ZA	zygomaxillare - inferior malar	X2	13.2		15.5		12.2			12
15	LO	lateral orbit	X2	8.1	4.7						11.5

		Facial Points		Average (mean)	Japanese Female	Chinese (NZ) F- 25-30yrs	Chinese Female	Korean Female	Korean Female	Buryats Female	Asian-derived - SW Am. Indian - N Female
16	ZY	lateral zygomatic - mid	X2	5.2	2.9			5.6			7
17	SG	zygomatic - supraglenoid	X2	6.3							6.25
18	GO	gonion	X2	9.1	4.0	10.8	14.72	5.4			10.5
19	1M	supra M2 molar	X2	17.2	9.7	31.7		9.3			18
20	MM	midmasseter - occlusal line	X2	15.9		13.1		17			17.5
21	2M	sub M2 molar	X2	18.3	12.3	29.2		14.6			17
		Source - Helmer		Age Unknown	unattributed internet source - http://what-when-how.com/forensic-sciences/facial-tissue-thickness-in-facial-reconstruction/	L Baillie, SA Mirijali, B Niven, et al (2015)	Aulsebrook WA1, Becker PJ, Işcan MY (1996)	from Forensic Facial Reconstruction by Caroline Wilkinson	Aulsebrook WA1, Becker PJ, Işcan MY (1996)	Aulsebrook WA1, Becker PJ, Işcan MY (1996)	Rhine (1983)

✳ All tissue depth measurements are in millimetres.

Reconstruction process

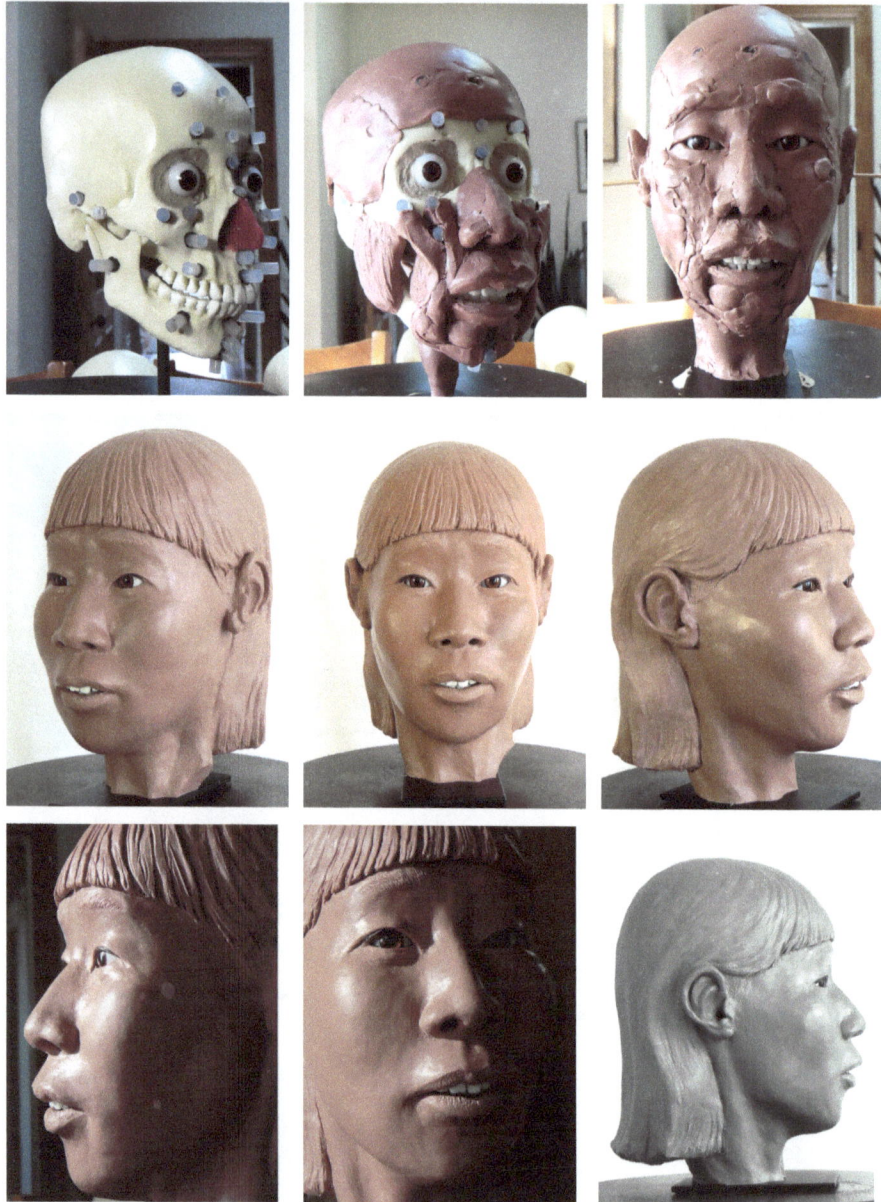

* I depicted subject younger than indicated in the Osteological Evaluation.

Asian male skull

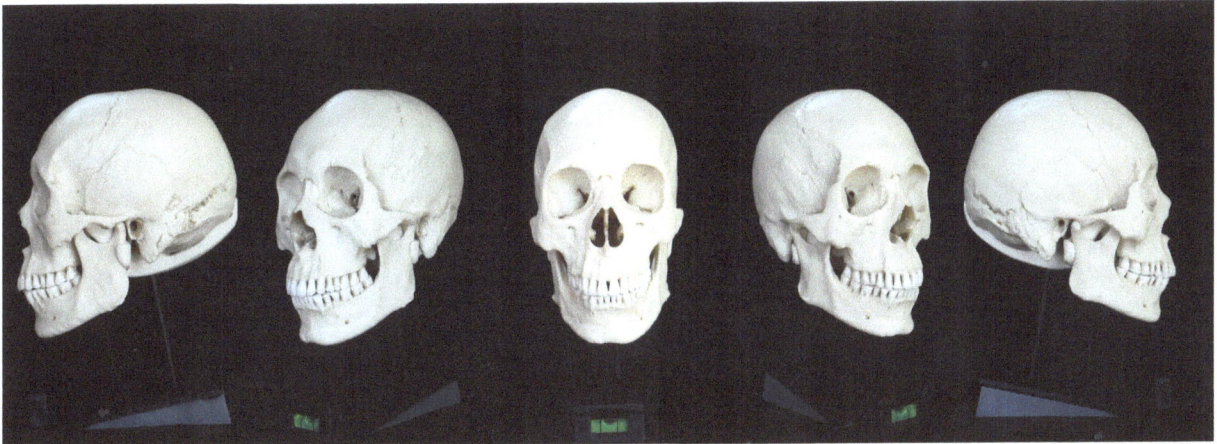

The morphology of the Asian male skull is distinct from that of other racial groups. Some of the most notable features include a wider and flatter mid-face, a more prominent brow ridge, and a shorter nasal bone. The zygomatic arches, or cheekbones, tend to be wider and flatter, and the mandible or jaw bone may be more robust. Genetic and environmental factors influence these features and may vary to some extent between Asian populations.

One of the key factors contributing to the unique morphology of the Asian male skull is the presence of a prominent supraorbital ridge or brow ridge. This bony ridge extends horizontally across the forehead above the eyes and can be quite pronounced in some individuals. It may play a role in determining the overall shape and contour of the face, giving it a more angular and masculine appearance.

While considerable variability exists within and between different Asian populations, certain features tend to be shared across this ancestral group.

ASIAN MALE SKULL

Features	Pre-construction observations and indicators - skull observed in the Frankfort Plan. Additional information from Bone Clone's report are added in Red. My additional comments may be added to notes (in bold) after reconstruction.
General Observations	Asian male - large and robust looking. This individual appears to be <50yrs, based on (my limited knowledge of) dentition. Though a controversial method, there is little fusion of the coronal, lambdoid, and sagittal sutures. This individual displays Asian feature, with the more rounded orbital orifice. The nasal opening is an upside down heart shape, with a projection of bone at the end of the nasion. This individual seems fairly symmetrical, with the exception of a slightly heavier right-side brow ridge, and eye-socket. Whilst the right-side zygomatic bone is slightly heavier, it is more projecting when viewed from above. It is noted that both styloid process are present.
OVERALL SKULL SHAPE	The distance from the glabella to the external occipital protuberance is approx 174mm. The distance widest breadth is approx 125mm, with a cephalic index of 71.8.* This index makes for a **dolichocephalic** skull. In profile, from the glabella to the vertex is approx. 121mm. The skull protrudes at the top of the frontal bone and dips slightly after the coronal suture (slight post-bregmatic depression). The cranial vault curves down from the lateral view, with the external orbital protuberance being the most posterior point of the skull. The superior and inferior temporal linea are strongly defined. Strong external occipital brow ridges are evident. From the posterior view, the cranial vault is somewhat 'squarish' - sloping from the sagittal suture down the parietal bones where from the midpoint, they fall vertically, nearly straight down to the mastoid processes. There are two anomalous foramen in the back of the skull which fall along the line of the sagittal suture and above the area of the posterior fontanelle. There is a 'triangle' of bone at the junction of the sagittal suture and the lambdoid sutures. The external occipital protuberance measures about 55mm to the posterior edge of the foramen magnum.

FOREHEAD	From the glabella to the vertex is approx 120mm. From the vertex to the most posterior point of the skull is approximately another 130mms. From the glabella, the forehead sloped back slightly, then curves back to the highest point on the frontal bone. Noting the change from 'smooth' skull to a more grainy texture, I estimate that the hairline starts approx 60mm from the nasion (measured with callipers).
BROW SHAPE	Strong brow ridge. Bony structure medial to the mid-supraorbital point, which is stronger on the right side. This may indicate a more 'raised' eyebrow on the left side. The flatness of the orbital opening at the top of the brow suggest the eyebrows will be superficial and straight. The attachments for the Corrugator Spericilii and Procerus muscle groups appear to be prominent, suggesting that there may be either/or both forehead, mid-nasal creases. Note that the R.Supraorbital Notch (foramen) hasn't formed into a typical foramen 'hole', or has broken either peri or postmortem, or missing in the casting process.
EYE FISSURE	'Rounder' shaped orbits, with a angular lower margins. Left orbit is slightly larger overall than the right. The horizontal measurement for the right side opening is slightly larger than the left. (V-R37mm V-L34mm, H-R36mm, H-L34mm)
EYEBALL	The eyeball sockets are quite shallow compared to the African and European subjects. Using a 25mm ball as a guide, it appears as though the eyes will be 'protruding' somewhat within the orbit cavity. It may also mean I use a smaller eyeball size during reconstruction. The plane of the orbit cavities appear even. Lining up the lacrimal crest and malar tubercle, the left eye will be slightly more raised than the left side. The interocular distance is broad.
EYELID PATTERN	With a thicker lateral rim on the left side, the eyelid may 'dip' more on that side. Given the ancestry of this skull, I expect to produce a recognisable epicanthal (Mongolian) fold that will semi-obscure the lacrimal caruncle and be determined by the projection of the nasion.
EYEBROW PATTERN	Eyebrows will be slightly tucked under the ridge, then follow the brow line. See photo.

NASAL PROFILE	Nasal projection is somewhat flat and shallow, with a small rounding at the end of the nasion. The nasal spine itself measures approx 7mm from volmer to tip (though the tip is not well defined). A total of 27.5mm projection will be added to the philtrum depth of approx 10.1mm.** However, given ancestry, I will use alternate calculations (methods) and adjust the dimensions to fit the face, should the nose projection look out of place.
NASAL WIDTH	The overall shape of the opening is heart shape with a bony overhang at the end of nasion. The overall width, taken at the nasion, just above the lacrimal crest, is 23mm. The nasal aperture (piriform aperture) is 25.3mm at its widest, indicating overall width between 42-46mm - taking into consideration ancestry. The septum is straight, and there is some nasal guttering. The base of the nasal opening is even, but slightly wider and rounder on the lateral right side.
NASAL SPINE	The nasal spine is well defined and straight. The spine itself angles downward. The nasal spine itself measures approx 7mm from volmer to tip.
ALAR SHAPE AND POSITION	Alar position will be slightly more rounded on right side. Small alar 'shelf' or nasal guttering.
NOSTRIL POSITION	The nasal position may not be evident (in the Frankfort plane) once tissue has been added
TEETH	Overall, the teeth seem be in good condition, with little wear, and appear to be 'average' in size.*** All 32 teeth are in the skull. All wisdom teeth have erupted. Wear is evident on the canines, and in both 1Ms, with some on both 2Ms, and little on the 3Ms. I would estimate age of 30-45yrs. Cranial vault (landmarks 1-7) 39.4 +/-9.1yrs. Anterior cranium (landmarks 6-10) 45.5 +/-8.9yrs). Estimate 36.6yrs to 48.5yrs. The teeth show 'shovelling' - the posterior/interior of the incisors are 'scooped out' . The bony structure of the mandible is receded around the lower canines and 1st bi-cuspids, leaving the root exposed (severe attrition - there is mild to moderate buccal furcation involvement of the mandibular 1st molar and the maxillary 1st & 2nd molar teeth). This could indicate that this individual had gum disease, or may have been a smoker (see photo). The dental arcade is somewhat rounded. There is mild alveolar prognathism.

LIP THICKNESS	The upper (maxillary) and lower (mandibular) teeth show no prognathism. The occlusion line is in a slight 'smile'. I would expect flat lips, with the appearance of more than average space between the end of nasion and the top of the lip (27+mm from end of nasal spine to occasional line/mid central incisor).
PHILTRUM WIDTH	Approx 10mm - Incisor angle suggest that the philtrum may be more rectangular, as opposed to rectangular.
MOUTH WIDTH	Approx 37mm - measured from mid-incisors.
MOUTH CORNER INCLINATIONS	The markings for the levator anguli oris and the depressor anguli oris seem prominent. I would expect the corners to be horizontal/neutral.
LIP SHAPE	The upper enamel line is relatively horizontal. Given the African ancestry, the lips may be full. With the bony spine running vertically down the philtrum, meeting the vermillion line, might make the upper lip seem more turned up.
NASOLABIAL FOLD	Moderately shallow canine fossa - making for a 'flatter', more rounded face.
CHIN SHAPE	The chin has little mental eminence, but is strong and square on either side of the midline. The mandible is robust, symetrical, and the angle is approx 110degs, with strong attachments and significant flair at the gonion. The inferior border of the mandible is somewhat squared.
EAR	The supra mastoid crest is smooth and flat. The mastoid process is cannot be seen when viewing face on, indicating the ear may not be easily visible. The mastoid process is pointing forward slightly, indicating that the earlobe may be attached.**** The ear will be angled to reflect the 110degs of the mandible.
CHEEK SHAPE	The zygomatic arches are robust, with a breadth of 128mm - 3mm wider than the breadth of the cranium. The right side is more prominent that the left. The superior line of the zygomatic arch is fairly straight and with a zigzag contour underneath (suggesting strong masseter attachments), as it merges toward the temporal junction.
References:	

* Cephalic Index used to determine dolichocephalic (long headed), mesaticephalic (moderate headed), or brachycephalic (short headed) cephalic index or cranial index. Anders Retzius (1796–1860) An index < 75 - **dolichocephalic**. An index of 75-80 - **mesaticephalic**. An index >80 - **brachycephalic**.

** Width - (European): aperture of 3/5ths the overall nose (Gerasimov). (Negroid): aperture measured at its widest point; 16mm added for total width (8mm added to each side of aperture), (Krogman - Karen Taylor). **Alternately**: The profile of the nose as determined by the Lebedinskaya method. Line A dissects the nasion and prothion. Line B is parallel to line A and intersects the foremost point on the nasal bone. Four to six equidistant lines of the piriform aperture. At each perpendicular line, the distance from line B to the piriform rim was measured and the same distance added on the other side of the line B. Using this method the profile of the nose was predicted. modified from Prokopec and Ubelaker (2002).

*** Dr Welfel (1974-1974) study of tooth size - averages

**** Fedosyutkin and Nainys (1993) estimates of ear positioning.

Tissue depths

Asian Male (mean)

		Facial Points		Average (mean)	Japanese Male	Korean Male	Chinese Male	Korean Male	Buryats Male	Asian-derived - SW Am. Indian - N Male
1	O	supraglabella		4.3	3.0	5.2	3.98			5
2	G	glabella		5.1	3.8	5.1	5.43		5.4	5.75
3	N	nasion		5.0	4.1	4.5		4.5	4.8	6.86
4	RHI	end of nasal		2.8	2.2	2.8	2.65			3.5
5	MP	mid-philtrum		10.9		11.1	11.85			9.75
6	LS	upper lip margin		12.1		12.6		12.6	13.5	9.75
7	LI	lower lip margin		13.3		13.8		13.8	14.5	11
8	LM	labiomental - lip fold		11.1	10.5	11.3				11.5
9	M	mental eminence - pogonion		10.0	6.2	10.6	9.42	10.6	11.4	12
10	MN	(menton - beneath chin) - gnathion		6.2	4.8	6.3	5.57			8
11	FE	lateral forehead	X 2	4.4		4.5				4.25
12	OS	(mid)-supraorbital	X 2	7.5			5.95			9
13	OR	orbitale - sub	X 2	5.6	3.7					7.5
14	ZA	zygomaxillare - inferior malar	X 2	11.9		9.8				14
15	LO	lateral orbit	X 2							12.5

		Facial Points		Average (mean)	Japanese Male	Korean Male	Chinese Male	Korean Male	Buryats Male	Asian-derived - SW Am. Indian - N Male
16	ZY	lateral zygomatic - mid	X2	5.5	4.4	4.7				7.5
17	SG	zygomatic - supraglenoid	X2							8.5
18	GO	gonion	X2	9.9	6.8	4.6	14.98			13.25
19	1M	supra M2 molar	X2	14.0	10.2	10.4				21.5
20	MM	midmasseter - occlusal line	X2	18.9		17				20.75
21	2M	sub M2 molar	X2	15.5	14.5	12.8				19.25
		Source - Helmer		Age Unknown	unattributed internet source - http://what-when-how.com/forensic-sciences/facial-tissue-thickness-in-facial-reconstruction/	Forensic Facial Reconstruction by Caroline Wilkinson	Aulsebrook WA1, Becker PJ, Işcan MY (1996)	Aulsebrook WA1, Becker PJ, Işcan MY (1996)	Aulsebrook WA1, Becker PJ, Işcan MY (1996)	Rhine (1983)

* All tissue depth measurements are in millimetres.

Reconstruction process

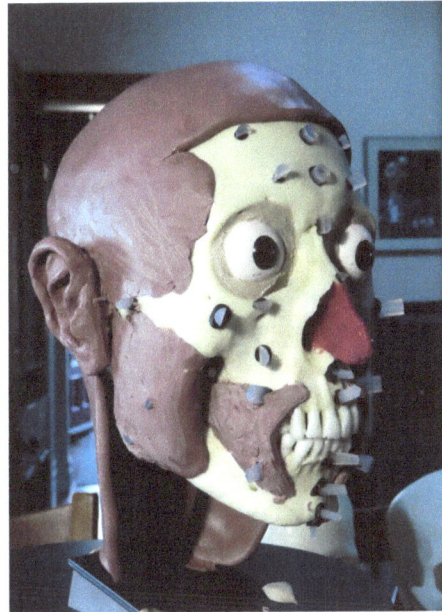

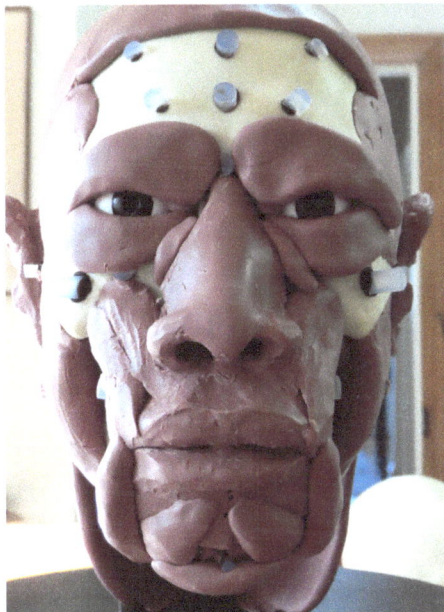

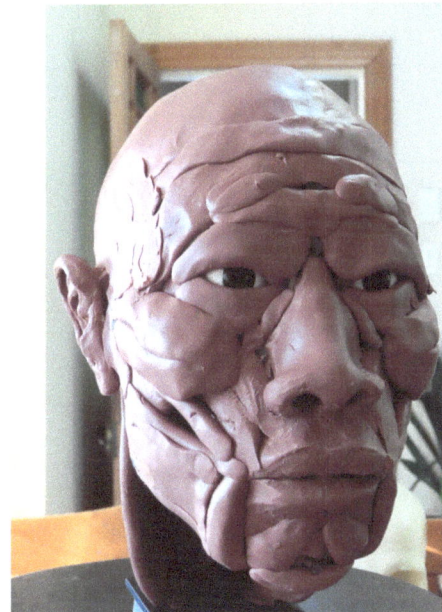

reconstruction process cont...

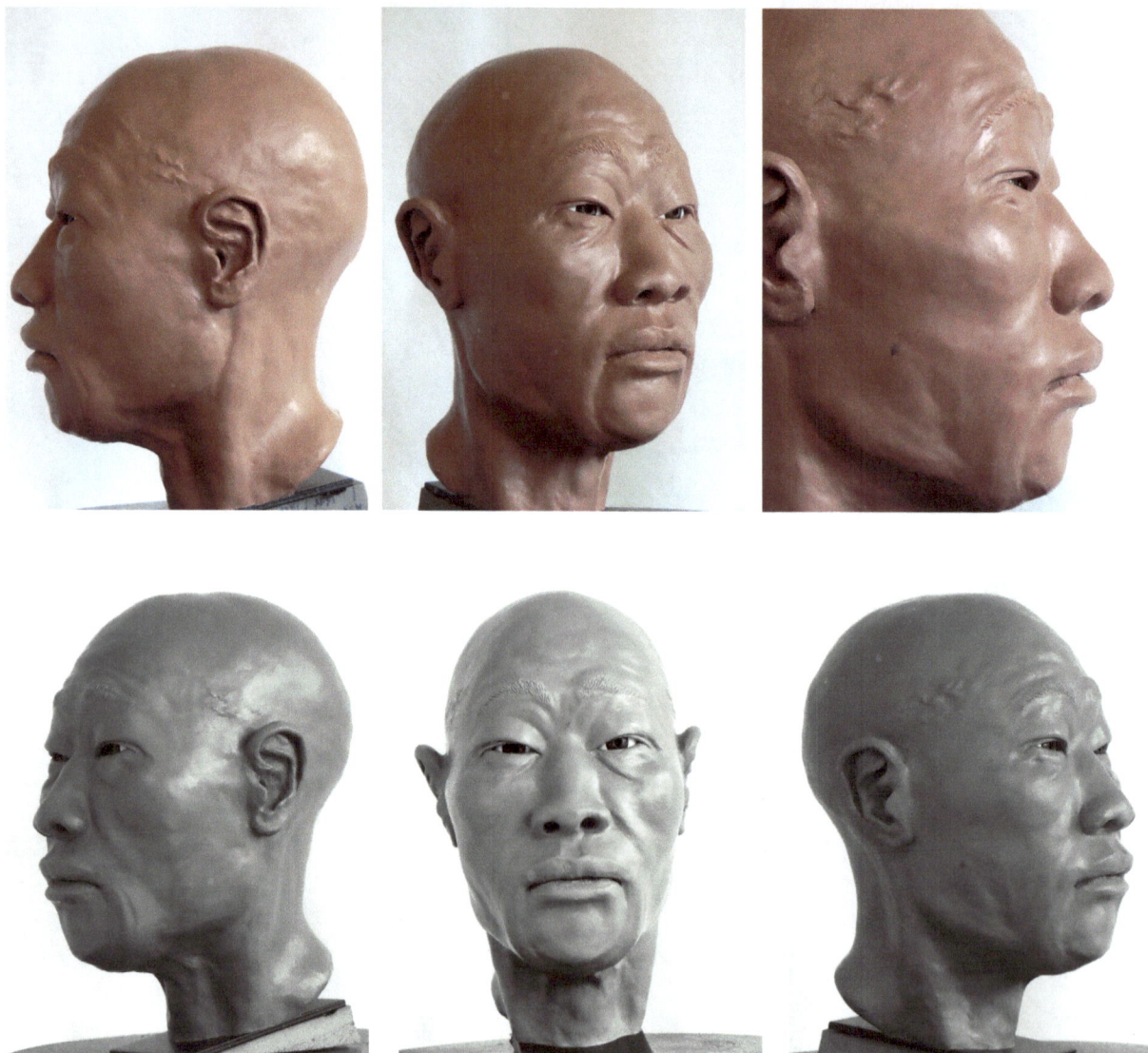

European female skull

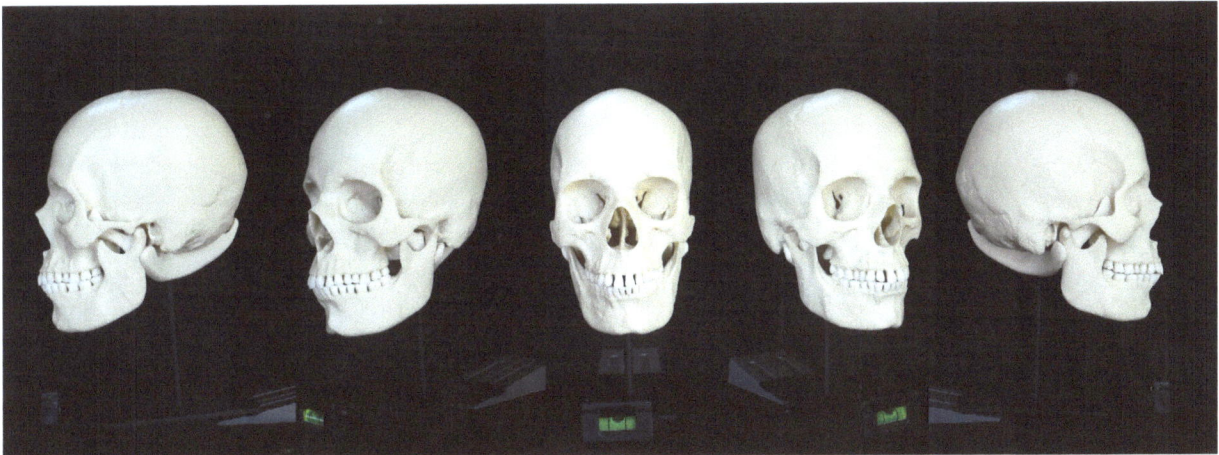

One of the most notable features about the European female skull is the relatively small size of the cranium, especially when compared to males of the same population. This difference in size is more pronounced in the facial region, where the female skull tends to be narrower and more gracile than the male skull. The forehead is typically more vertical and less sloping, while the occipital region tends to be flatter.

Another feature of the European female skull is the shape of the orbits, or eye sockets. These are generally more rounded and forward-facing than in other populations, giving the face an open and expressive appearance. The nasal bones are also typically narrower and more delicate, with a more defined bridge than in some other populations.

This particular female skull impressed me with its robust and square jawline and I wondered about the variation that exists between different populations and ancestral groups.

For example, studies have shown that females from Northern Europe tend to have narrower faces and more prominent cheekbones than those from Southern Europe. Similarly, females from Eastern Europe tend to have larger and more prominent jaws than those from Western Europe.

EUROPEAN FEMALE SKULL

Feature	Pre-construction observations and indicators - skull observed in the Frankfort Plan. Additional information from Bone Clone's report are added in Red. My additional comments may be added to notes (in bold) after reconstruction.
General Observations	European Female. This individual appears to be in very good condition and <50yrs, based on (my limited knowledge of) dentition (loss of teeth and remodelled bone - premortem and postmortem) and (though a controversial method). There is no total fusion of the coronal, lambdoid, and sagittal sutures. This individual displays typical European features, with the more 'aviator' shaped orbital orifices, and a 'triangular' nasal opening. This individual is very symmetrical, with the exception of a heavier right-side of the mandible. Whilst the right-side zygomatic bone is heavier, it equal to the left side when viewed from above. Both styloid process are present. Suggestion of a small sutra bone (Wormian ossicle) at the right asterion).
OVERALL SKULL SHAPE	The distance from the glabella to the most posterior point at the occipital bone, just below the posterior fontanelle, is approx 165mm. The distance widest breadth is approx 124mm, with a cephalic index of 75.2* This index makes for a **dolichocephalic** skull. In profile, from the glabella to the vertex is approx. 113mm and 115mm from the vertex to the most posterior point on the skull. The skull dips slightly after the coronal suture (post-bregmatic depression). The cranial vault is nicely rounded from the lateral view. From the posterior view, the cranial vault is very rounded to mid-parietal bones, then nearly straight down to the mastoid processes. The external occipital protuberance measures about 36mm to the posterior edge of the foramen magnum.
FOREHEAD	From the glabella to the vertex is approx 100mm. From the glabella to the vertex, the skull is very smooth, flat, and verticle, before sloping back to the coronal suture. Noting the change from 'smooth' skull to a more grainy texture, I estimate that the hairline starts approx 60mm from the nasion (measured with callipers).

BROW SHAPE	Flat vertical brow ridge (sideview). Very slight bony structures medial to the mid-supraorbital point, but no suggestion that the brow would be overhang the orbit. The attachments for the Corrugator Spericilii and Procerus muscle groups do not appear to be prominent, suggesting that she may have had a smooth brow.
EYE FISSURE	'Aviator' shaped orbits, with a curved lower margin. The orbits are very symmetrical. The horizontal measurement for the right side opening is 1mm larger than the right, and equal in the vertical. (V-R32mm V-L32mm, H-R34 H-L35.5mm).
EYEBALL	Using a 25mm ball as a guide, it appears as though the eyes will be 'average' depth within the orbit cavity, with the right eye being slightly forward of the left. However, given a deep eye-socket and posterior positioning of the malar tubercle, the eyes may have a more sunken appearance. Lining up the lacrimal crest and malar tubercle, the eyes are very 'straight' (only slight lateral angle). The right lacrimal crest is slightly posterior to the left.
EYELID PATTERN	With a straight superior rim, it is estimated that the eyelid fold will be heavier on top and straight. L.eye seems to have slightly more med-supraorbital overhang. The left eye may 'droop' more in the centre.
EYEBROW PATTERN	Eyebrows will be on the surface of the brow ridge and slightly curved. See example photo.
NASAL PROFILE	Nasal projection is pronounced, indicative of her European ancestry. The nasion is straight, then has a slight downward curve. The nasal spine itself is not well defined, but seems to measures 4mm from volmer to tip. A total of 12mm projection will be added to the philtrum depth of approx 8.5mm (20.5mm), which may be too short.** Various calculations (methods) and adjust the dimensions to fit the face, should the nose projection look out of place. Nasal root is prominent and the nasal angle is acute.
NASAL WIDTH	The overall shape of the opening is an elongated, triangular shape. The overall width, taken at the nasion, just above the lacrimal crest, is 22.8mm. The nasal aperture (piriform aperture) is 25mm at its widest, indicating overall width between 41.6mm. The the nasal opening is slightly wider on on the right side and slightly larger.

NASAL SPINE	The nasal spine is not well defined. The spine itself is slightly downturned, whist the underside points 'straight' forward. The underside of the nasal spine is blunt, then carries on downward with a vertical thin column/spine of bone running down the maxilla/philtrum the a projection of approx 3.5mms before disappearing into the bone above the central incisors (see photo below). The nasal spine itself measures 7.5mm from volmer to tip.
ALAR SHAPE AND POSITION	Alar position will be slightly higher and slightly wider on right side. No notable alar 'shelf' or nasal guttering.
NOSTRIL POSITION	The nasal position is expected to be evident (in the Frankfort plane) once tissue has been added
TEETH	Overall, the teeth seem be in good condition with little wear, and appear to be slightly smaller than 'average' in size.*** All 32 teeth are present. All erupted wisdom teeth are present. The top 4 incisors have more than average spacing/gaps, and both central incisors are angled up medially (perhaps 'chipped'- see photo). The R central and lateral incisors are protruding and do not seen to make contact on the occasional line. The lack of wear either denotes a younger age at death, or perhaps a softer 'western' diet. The occlusion line is 'wave-like'. I would guess age to be 25-40yrs. Crainial vault (landmarks 1-7) 39.4yrs +/- 9.1yrs. Anterior cranium (landmarks 6-10) 56.2 +/- 8.5yrs. Age range 30.3-64.7 years. Moderate degree of attrition. Dental arcade is somewhat V-shaped.
LIP THICKNESS	The R central and lateral incisors are protruding and do not seen to make contact on the occasional line and may make the right side of the lips protrude slightly more than the left side. Given the smaller than average teeth, both the upper and lower lips may be on the thinner side.
PHILTRUM WIDTH	Approx 10mm - Incisor angle suggest that the philtrum may be more square.
MOUTH WIDTH	Approx 36mm - measured from incisors.

MOUTH CORNER INCLINATION S	The markings for the levator anguli oris seem more pronounced than the depressor anguli oris, indicating that mouth corners may turn up instead of down. However, age may play a part in more downturned corners (marionette lines).
LIP SHAPE	The upper enamel line is slightly turned upwards. The lips may be on the thinner side.
NASOLABIAL FOLD	Fairly deep canine fossa - could indicate nasolabial fold if the individual was slim.
CHIN SHAPE	The chin has a strong mental eminence, with the bone higher in the middle, then lower on either side. Perhaps this might indicate a dimple on the chin (see quarter view photo). The mandible is markedly robust (for a female) and rounded, with an angle is 115degs, with weak muscle attachments at the gonion.
EAR	The supra mastoid crest is smooth and straight. The mastoid process is rougher and can not be seen when viewing face on. The ears may be quite flat against the head. The mastoid process is pointing downwards, indicating that the earlobe may be attached.**** The ear will be angled to reflect the 115degs of the mandible.
CHEEK SHAPE	The zygomatic arches appear short, and are not significantly robust. The overall breadth is 120mm, 4mms less than that of the cranium. The zygomatic arch itself is smooth and curved upwards as it merges toward the temporal junction. The right side of the zygomatic bone is slightly more robust - than the left side.
References:	
	* Cephalic Index used to determine dolichocephalic (long headed), mesaticephalic (moderate headed), or brachycephalic (short headed) cephalic index or cranial index. Anders Retzius (1796-1860) An index < 75 - **dolichocephalic**. An index of 75-80 - **mesaticephalic**. An index >80 - **brachycephalic**.

** Width - (European): aperture of 3/5ths the overall nose (Gerasimov). (Negroid): aperture measured at its widest point; 16mm added for total width (8mm added to each side of aperture), (Krogman - Karen Taylor). **Alternately**: The profile of the nose as determined by the Lebedinskaya method. Line A dissects the nasion and prothion. Line B is parallel to line A and intersects the foremost point on the nasal bone. Four to six equidistant lines of the piriform aperture. At each perpendicular line, the distance from line B to the piriform rim was measured and the same distance added on the other side of the line B. Using this method the profile of the nose was predicted. modified from Prokopec and Ubelaker (2002).

*** Dr Welfel (1974-1974) study of tooth size - averages

**** Fedosyutkin and Nainys (1993) estimates of ear positioning.

Tissue depths

Caucasoid / European Female (mean)

		Facial Points		Average (Mean)	Fem. Amer/ Caucasoid	Caucasoid Female	40yr+ Amer European (F)	
1	O	supraglabella		3.5	3.5	5.2	5.2	
2	G	glabella		4.8	4.75	5.9	5.9	
3	N	nasion		5.5	5.5	5.5	6.2	
4	RHI	end of nasal		2.8	2.75	2.7	2.4	
5	MP	mid-philtrum		8.5	8.5	8.5	12.6	
6	LS	upper lip margin		9.0	9	9	10.5	
7	LI	lower lip margin		10.0	10	10	12.5	
8	LM	labiomental - lip fold		9.5	9.5	9.5	12.3	
9	M	mental eminence - pogonion		10.0	10	10	9.6	
10	MN	(menton - beneath chin) - gnathion		5.8	5.75	5.7	6.9	
11	FE	lateral forehead	X 2	3.5	3.5	3.5	3.5	
12	OS	(mid)-supraorbital	X 2	7.0	7	7	7.4	
13	OR	orbitale - sub	X 2	6.0	6	6	5.4	
14	ZA	zygomaxillare - inferior malar	X 2	12.8	12.75	12.7	10.6	
15	LO	lateral orbit	X 2	10.8	10.75	7.4	5.4	

143

		Facial Points		Average (Mean)	Fem. Amer/ Caucaso id	Caucasoi d Female	40yr+ Amer European (F)	
16	ZY	lateral zygomatic - mid	X 2	7.5	7.5	7.5	9.1	
17	SG	zygomatic - supraglenoid	X 2	8.0	8	8	6.2	
18	GO	gonion	X 2	12.0	12	12	11.2	
19	1M	supra M2 molar	X 2	19.2	19.2	19.25	20.5	
20	MM	midmasseter - occlusal line	X 2	17.0	17	17	17.8	
21	2M	sub M2 molar	X 2	15.5	15.50	15.5	18	
		Source		Unknown Age	Rhine & Moore	Kollyman Buchly 1898	compilation (Wilkinson)	

✳ All tissue depth measurements are in millimetres.

144

Reconstruction process

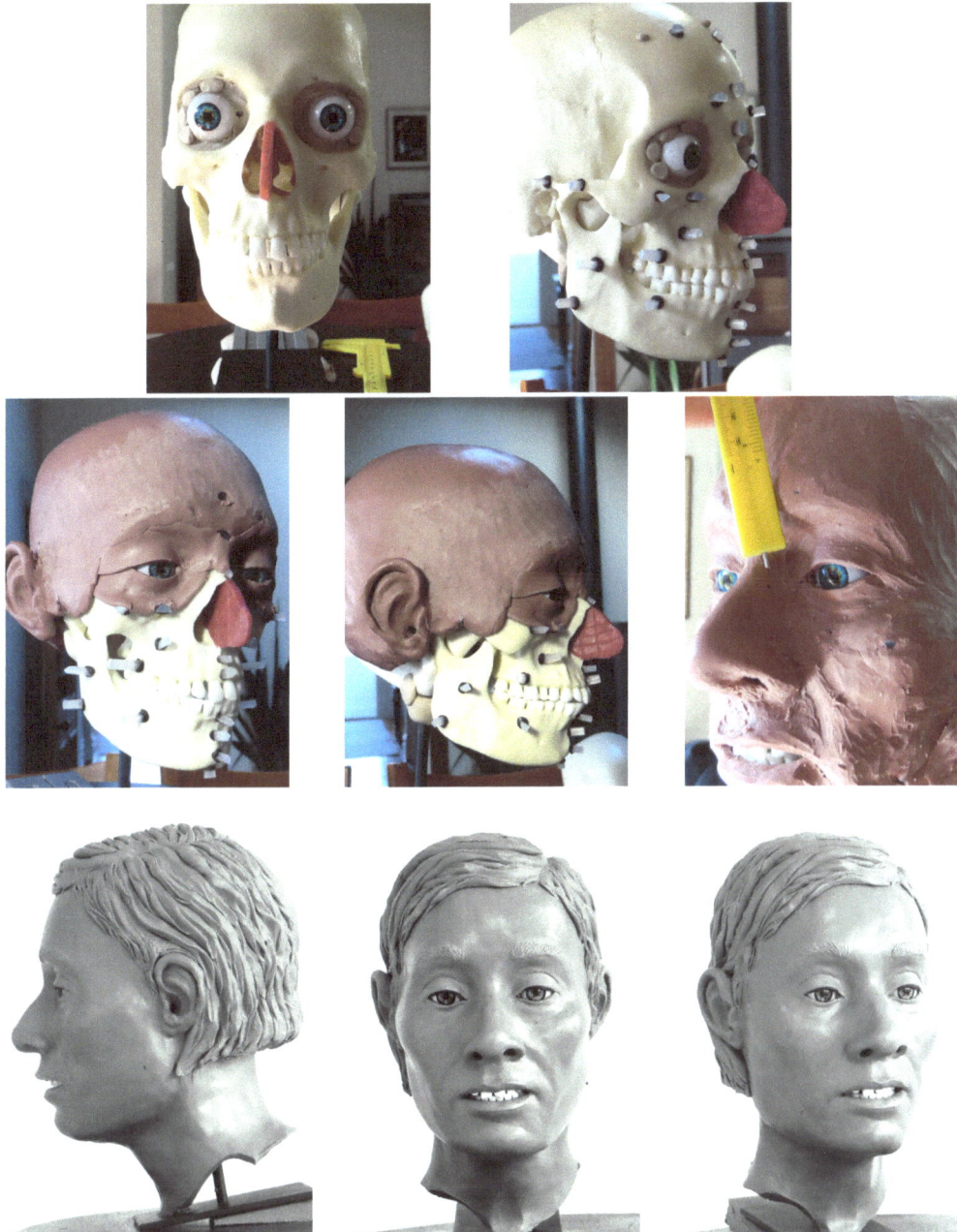

European male skull

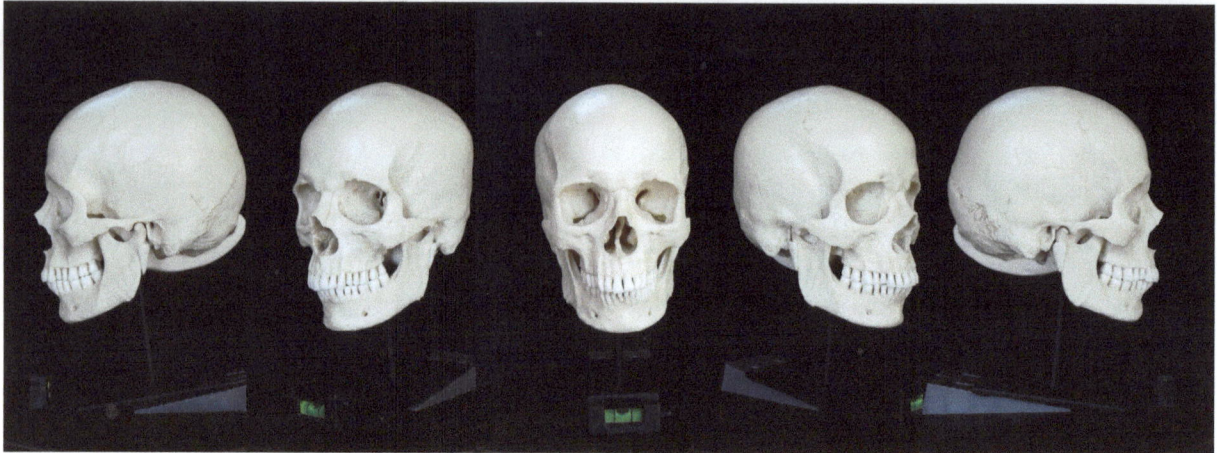

Generally, the European male skull is characterised by a broad and high forehead, a relatively long and narrow face, and a pronounced brow ridge. The skull is also generally symmetrical, with well-defined cranial sutures.

The prominent supraorbital ridge, or brow ridge, is a notable feature of the European male skull. The degree of brow ridge development can vary among individuals and is often used as a metric in forensic anthropology to determine the gender of skeletal remains.

Another feature of the European male skull is the nasal index, which is the ratio of the width of the nasal aperture to its height. This index tends to be relatively low in European males, indicating a narrower and more elongated nose compared to other populations. Additionally, the maxillary bone, which forms the upper jaw, is typically longer in European males, resulting in a narrower mid-face region.

EUROPEAN MALE SKULL

Feature	
	Pre-construction observations and indicators - skull observed in the Frankfort Plan. Additional information from Bone Clone's report are added in Red. My additional comments may be added to notes (in bold) after reconstruction.
General Observations	European Male. This individual could be >50yrs, based on (my limited knowledge of) dentition and remodelled bone (antemortem and postmortem). Though a controversial method, there is no total fusion of the coronal, lambdoid, and sagittal sutures. This individual displays typical European features, with the more 'aviator' shaped orbital orifices, and a 'teardrop' shaped nasal opening. It is apparent that this individual suffered facial trauma antemortem, as there is evidence of a very deviated septum/nasal spine. The skull, though complete has many obvious 'fracture' lines through the orbital cavity, frontal/zygomatic/maxilla bone junctures, including a fracture line running vertically from the nasal spine through the upper central incisors. It is unclear whether these 'fractures' are representative of the individual or artefacts of the casting process (see photo). Other than the evident of trauma to the nose, this individual is fairy symmetrical. Both styloid process are present. Cranial vault age estimates: landmarks 1-7 (score of 14) 45.2 +/-12.6yrs. Landmarks 6-10 (score of 10) 51.9 +/- 12.5yrs. Estimate age 39.4 - 57.8yrs.
OVERALL SKULL SHAPE	The distance from the glabella to the most posterior point at the occipital bone, just below the posterior fontanelle, is approx 170mm. The distance widest breadth is approx 134mm, with a cephalic index of 78.8* This index makes for a **mesaticephalic** skull. In profile, from the glabella to the vertex is approx. 115mm and 115mm from the vertex to the most posterior point on the skull. The skull continues to curve up after the coronal suture. The cranial vault is nicely rounded from the lateral view. From the posterior view, there is a more prominent 'peak' at the sagittal suture. The cranial vault is 'triangular to mid-parietal bones, where they are most prominent, and higher on the right side. From there, the parietal bones curve, slightly down to the mastoid processes. The external occipital protuberance has prominent ridge and measures about 42mm to the posterior edge of the foramen magnum. Two small sterol bones (Wormian ossicles) at both right and left asterions.

FOREHEAD	From the glabella to the vertex is approx 115mm. The forehead is very vertical before arching back toward the digital suture. The frontal bone, from the forehead, is smooth and flat before meeting the coronal suture. Noting the change from 'smooth' skull to a more grainy texture, I estimate that the hairline starts approx 85mm from the nasion (measured with callipers). This may indicate the individual had a receding hairline (or was perhaps bald).
BROW SHAPE	The brow ridge is fairly strong, particularly either side of the glabella extending to the mid-supraorbital point. The attachments for the Corrugator Spericilii and Procerus muscle groups appear to be prominent, with two notable 'protuberances' either side of the nasion (supraorbital tori). This may suggest that this individual may have had vertical 'frown' lined between his brow, drawing the brow closer together medially. The brow suggestions that the brow would be overhang the orbit, then follow the sharp supraorbital rim in a 'straight' line. See photo example below. Note that both Supraorbital Notch (foramen) are indicated but haven't formed into a typical foramen 'hole'. It is unclear whether this is present in the skull or an artefact from the casting process. The supraorbital margins are blunted.
EYE FISSURE	'Aviator' shaped orbits, with a curved lower margin. The orbits are very symmetrical. (V-R32mm V-L32mm, H-R36, H-L36mm).
EYEBALL	Using a 25mm ball within the orbit cavity as a guide, and given the posterior positioning of the zygomatic/frontal bone suture and the heavy brow ridge at the nasion, the eyes may appear 'deepset', having a more sunken appearance. Lining up the lacrimal crest and malar tubercle, the eyes are very 'straight' (only slight lateral angles).
EYELID PATTERN	With a straight superior rim on the right side, it is estimated that the eyelid folds will be heavier on top and straight. Considering the small orbit opening, I expect the eyelids to be hidden by the upper lid 'overhang' L.eye seems to have slightly more supraorbital angle. The left eyelid may 'heavier' laterally
EYEBROW PATTERN	Eyebrows will be under the surface of the medial brow ridge, perhaps heavier centrally 'kitted brow', with some angulation upwards toward the lateral sides. See photo example.

NASAL PROFILE	Nasal projection is pronounced and narrow, indicative of European ancestry. The nasion has an upward curve, then turns down slightly at the end of nasion. The nasal spine veers to the left and measures 11mm from volmer to tip. A total of 33mm projection will be added to the philtrum depth of approx 13mm (totalling 46mm), which may be far too long.** Various calculations (methods) and adjust the dimensions to fit the face, should the nose projection look out of place. I expect the nose to be noticeably 'crooked', veering toward the left side of the face. The nasion is rough.
NASAL WIDTH	The overall shape of the opening is an elongated, triangular shape, and irregular. The nasal bone is narrow at the nasion and then widens as it progresses toward the end of nasion. There is indication of nasal trauma, with evident deviated septum and the nasal spine veering to the left. (See photo) The right nasal opening is higher and wider, and the maxilla at the side of the nose recedes approx 2mm. The overall width, taken at the nasion, just above the lacrimal crest, is 23.6mm. The nasal aperture (piriform aperture) is 25.8mm at its widest, indicating overall width of approx 43mm. The the nasal opening is slightly wider on on the right side and slightly larger.
NASAL SPINE	The nasal spine is well defined and sharp, and the external lateral view shows a slight turn upward at the end (short). The internal spine itself is slightly downturned, and from the volmer to the end of the spine is significantly deviated to the left. The underside of the nasal spine is pointed (though left 'leaning) and a fracture line runs vertically from the nasal spine through the upper central incisors. The nasal spine itself measures 11mm from volmer to tip. Nasal length appears to be approx 44mm (11X3+11.99mm of philtrum depth)
ALAR SHAPE AND POSITION	Alar position will be slightly higher and slightly wider on right side. Some notable alar 'shelf' or nasal guttering. The right alar will be wider, and posterior to the left.
NOSTRIL POSITION	The nasal position (in the Frankfort plane) may be slightly hidden once tissue has been added

149

TEETH	Overall, the teeth themselves seem be in good condition with little wear, and appear to be smaller than 'average' in size.*** All 32 teeth are present. All erupted wisdom teeth are present. There is indication of periodontal disease, as there is bone loss particularly around the cuspids and incisors, making this individual look 'long in the tooth', due to exposed root. If this individual was a smoker, it may also affect the overall skin texture (wrinkles). The top central right incisor is slightly lower than the left. There is some prognathism of the lower mandible. The lack of wear either denotes a younger age at death than estimated, or perhaps a softer 'western' diet. The occlusion line is 'wave-like'. Moderate degree of attrition.
LIP THICKNESS	Given the smaller than average teeth, both the upper and lower lips may be on the thinner side.
PHILTRUM WIDTH	Approx 8mm - Incisor angle suggest that the philtrum may be more square.
MOUTH WIDTH	Approx 34mm - measured from incisors.
MOUTH CORNER INCLINATIONS	The markings for the levator anguli oris seem less pronounced than the depressor anguli oris, indicating that mouth corners may turn down instead of up. Age may play a part adding pronounced marionette lines.
LIP SHAPE	The upper enamel line is straight. The lips may be on the thinner side.
NASOLABIAL FOLD	Fairly deep canine fossa - could indicate nasolabial fold if the individual was slim.
CHIN SHAPE	The chin is rounded, with no strong mental eminence, and symmetrical. The chin might be slightly receding. The mandible is curved, with an angle of 120degs. There are strong muscle attachments at the gonion and slight flaring. The inferior boarder of the mandible is somewhat square.

EAR	The supra mastoid crest is smooth and curved. The mastoid process is rougher and can just be seen when viewing face on. The ears may stick out, particularly at the bottom, from the head. The mastoid process is pointing downwards, indicating that the earlobe may be attached.**** The ear will be angled to reflect the 120degs of the mandible.
CHEEK SHAPE	The zygomatic arches appear average. The overall breadth is 123mm, 11mm less than that of the cranium. The zygomatic arch itself is slightly undulating and the distal side curves upwards as it merges toward the temporal junction. The right side of the zygomatic bone is slightly more robust and higher than the left side. Moderate prominence of the masseteric tuber-sixties of the mandible.

References:

* Cephalic Index used to determine dolichocephalic (long headed), mesaticephalic (moderate headed), or brachycephalic (short headed) cephalic index or cranial index. Anders Retzius (1796-1860) An index < 75 - **dolichocephalic**. An index of 75-80 - **mesaticephalic**. An index >80 - **brachycephalic**.

** Width - (European): aperture of 3/5ths the overall nose (Gerasimov). (Negroid): aperture measured at its widest point; 16mm added for total width (8mm added to each side of aperture), (Krogman - Karen Taylor). **Alternately**: The profile of the nose as determined by the Lebedinskaya method. Line A dissects the nasion and prothion. Line B is parallel to line A and intersects the foremost point on the nasal bone. Four to six equidistant lines of the piriform aperture. At each perpendicular line, the distance from line B to the piriform rim was measured and the same distance added on the other side of the line B. Using this method the profile of the nose was predicted. modified from Prokopec and Ubelaker (2002).

*** Dr Welfel (1974-1974) study of tooth size - averages

**** Fedosyutkin and Nainys (1993) estimates of ear positioning.

151

An Artist's Guide to Forensic Facial Approximation

Tissue depths

Caucasoid / European Male (mean)

		Facial Points		AverageMale	20-29 yrs	30-39 yrs	40-49 yrs	50-59 yrs	60+years	White American 45-55yrs	American Caucasoid
1	O	supraglabella		4.83	4.9	4.6	5.2	5.2			4.25
2	G	glabella		5.91	5.7	6.2	6	5.9	6.3	6	5.25
3	N	nasion		7.20	8.2	7.3	6.8	7.3	7.1	7.2	6.5
4	RHI	end of nasal		2.53	2.3	2.5	2.7	2.8	2.6	1.8	3
5	MP	mid-philtrum		12.99	15.5	14.6	15.6	14.3	12.9	8	10
6	LS	upper lip margin		11.73	14	12.3	12.6	11.8	9.9		9.75
7	LI	lower lip margin		13.33	14.2	14.9	14.2	13	12.7		11
8	LM	labiomental - lip fold		12.22	12	12.1	13.3	13	12.8	11.6	10.75
9	M	mental eminence - pogonion		11.42	9.7	10.3	11.7	13.7	12.3	11	11.25
10	MN	(menton - beneath chin) - gnathion		8.35	7.5	8.3	9.5	9.8	8.9	7.2	7.25
11	FE	lateral forehead	X2	5.58	5.5	6	5.5	6	6.2		4.25
12	OS	(mid)-supraorbital	X2	7.42	7.3	7.3	7.2	7.5	6.7	7.7	8.25
13	OR	orbitale - sub	X2	5.51	5.2	5	5.8	5.5	5.5	5.8	5.75
14	ZA	zygomaxillare - inferior malar	X2	10.31	9.5	9.9	10.1	10.1	9		13.25
15	LO	lateral orbit	X2	6.27	5.3	5.2	5.8	5.7	5.6		10

		Facial Points		Average Male	20-29 yrs	30-39 yrs	40-49 yrs	50-59 yrs	60+years	White American 45-55yrs	American Caucasoid
16	ZY	lateral zygomatic - mid	X2	7.55	7.5	7.6	6.8	8	7.5	8.2	7.25
17	SG	zygomatic - supraglenoid	X2	5.79	5.3	5.3	5.5	5.5	5	5.4	8.5
18	GO	gonion	X2	11.76	9.2	10.1	10.2	12	10.3	19	11.5
19	1M	supra M2 molar	X2	20.84	20.2	22	21.7	22.3	18.8	21.4	19.5
20	MM	midmasseter - occlusal line	X2	20.04	19.2	21.3	20.4	20.5	20.6		18.25
21	2M	sub M2 molar	X2	17.53	19	18.5	18.3	18.3	17.2	15.4	16
		Source		unknown age	Helmer (1984)	Helmer (1984)	Helmer (1984)	Helmer (1984)		Manhein et al. (2000)	Rhine & Moore (1984)

✳ All tissue depth measurements are in millimetres.

153

Reconstruction progress

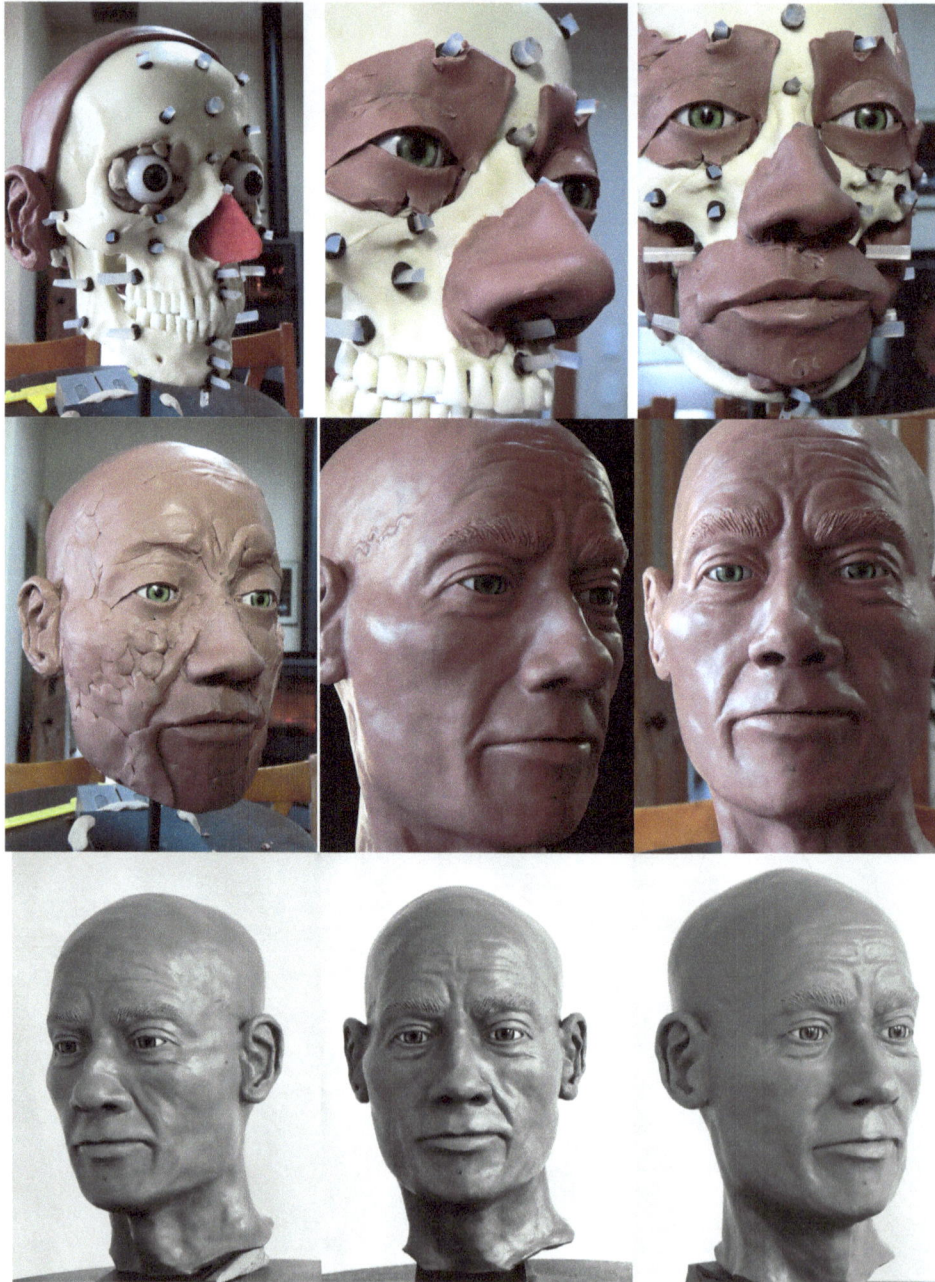

Bibliography

Wilkinson. Forensic Facial Reconstruction. 2004

John Prag and Richard Neave. Making Faces. 1999

K Taylor. Forensic Art and Illustration. 2001

Simon Mays. The Archaeology of Human Bones. 1998

Bruce and Young. In The Eye of the Beholder. 1998

Wilkinson and Ryan. Crainofacial Identification. 2012

Susan Hayes. 3D Facial Approximation Lab Manual. 2017

Eliot Goldfinger. Human Anatomy for Artists. 1991

Clement and Marks. Computer Graphic Facial Reconstruction. 2005

3Dtotal Publishing. Anatomy for Artists. 2020

Jenő Barcsay. Anatomy for the Artist. 2004 version

Gottfried Bammes. The Complete Guide to Anatomy for Artists & Illustrators. 1964

Uldis Zarins. Anatomy of Facial Expression. 2014

David K. Rubins. The Human Figure. 1953

Anthony Ryder. Figure Drawing. 2000

About the Author

Michele has been creating art since she could first hold a pencil. She drew pages and pages of fish, or so she was told. Apparently, fish were the simplest form for a small child to draw; an ellipse with a triangular tail. Her 'fish period' predates her memory.

Then Michele graduated to horses. Her love of horses blossomed at around three years of age after being strapped onto the back of a golden pony at the local pony rides. She remembers drawing horses endlessly. Michele drew on the front and back of every blank page of every book in the family home.

When Michele was around nine years old, her teachers saw her horse-drawing abilities, and she was asked to go to all the other elementary school classrooms to demonstrate to the other children how to paint a horse using thick, primary-coloured poster paint. Standing in front of those classes, particularly the ones with the older children, was an anxious moment for an introverted girl with social anxiety.

Throughout school, Michele was the kid who was always in the art room. It served as a relatively unstructured place of creativity and refuge, and she received the Art Award at high school graduation.

In the early 1970s, Michele decided to 'find her roots', exploring her French ancestry by counterintuitively attending the American College in Paris, where she studied drawing and art history.

In the early 1980s, Michele attended Northern Arizona University in Flagstaff, Arizona, USA., where she studied bronze casting and foundry, mould making, sculpture, anatomy for artists, skull and facial reconstruction, photography, mechanical drawing, and investigative/crime scene photography. There, Michele sculpted her first écorché figure (an anatomical figure built one bone at a time and then overlaid with each muscle.)

Because of this class, Michele became fascinated with the idea of being able to reconstruct a skull to identify its owner. However, it was also the advent of computer-aided facial reconstructions. This revolution in technology promised to make hands-on forensic facial reconstruction work redundant. That realisation, and a personal life change, discouraged her from pursuing forensic facial reconstruction as a career. She'd missed the boat, she decided.

Nevertheless, Michele continued bronze casting, sculpting Western art (horses, of course) in limited series, and received some commissioned figurative work. She also taught herself computer graphics and illustrated several books.

Moving to New Zealand in 1990, Michele produced a series of fish paintings (right back to her roots!). She even owned an art gallery for a time.

In 2006 Michele decided to up-skill her figurative work by attending an intensive four months of study at Studio Escalier in Argenton Chateau, France. She was particularly impressed with the Studio's unbroken lineage from teacher to student, dating back to one of her all-time favourite artists, Michelangelo.

At Studio Escalier, students strictly concentrate on classical figurative drawing and oil painting techniques in the traditional atelier style. It was demanding, but Michele acknowledges that she learned much on this intense course. The most important thing she learned, she says, is how to truly see what is there, not what one thinks is there.

Returning to New Zealand, Michele specialised in portraiture and figurative work, even being a finalist in New Zealand's Adams Portraiture Competition in 2012. More recently, she continued studying the human form with the very accomplished Spanish sculptor Javier Murcia, whose work, workshops, and online classes can be viewed at: www.workshopsculpting.com and https://javiermurcia.com/

But the allure of forensic facial reconstruction never really left her. After receiving a certificate in Forensic Facial Reconstruction from the University of Sheffield, Michele wanted more hands-on experience.

Purchasing beautifully cast replica skulls from ©Bone Clones, www.boneclones.com, and buying as many reference books as possible, Michele spent several years teaching herself what she could about forensic facial reconstruction. Now she wants to share her experience with other artists and hobbyists.

Michele is a current member of the Australian New Zealand Forensic Science Society.

www.ingramcontent.com/pod-product-compliance
Lightning Source LLC
Chambersburg PA
CBHW050917210326
41597CB00003B/125